Electron microscopy of soils and sediments: examples

Electron microscopy of soils and sediments: examples

PETER SMART and N. KEITH TOVEY

CLARENDON PRESS · OXFORD · 1981

Oxford University Press, Walton Street, Oxford OX2 6DP
London Glasgow New York Toronto
Delhi Bombay Calcutta Madras Karachi
Kuala Lumpur Singapore Hong Kong Tokyo
Nairobi Dar es Salaam Cape Town
Melbourne Wellington
and associate companies in
Beirut Berlin Ibadan Mexico City

Published in the United States
by Oxford University Press, New York

British Library Cataloguing in Publication Data

Smart, Peter
Electron microscopy of soils & sediments.
1. Soil science 2. Sediments (Geology) –
Analysis 3. Scanning electron microscope
I. Title II. Tovey, N. Keith
631.4'3 S591

ISBN 0–19–854515–0

Printed in Great Britain by
Butler & Tanner Ltd, Frome and London

Contents

List of figures

Similar examples in *Techniques*

Introduction

This collection of micrographs illustrates the structures of a variety of soils and sediments as seen in scanning and transmission electron microscopes. The experienced microscopist will be able to delve here at random; but the general reader is advised to read the captions in sequence, as the earlier ones instruct on the interpretation of the micrographs. The reader who is unfamiliar with these instruments is reminded that the sample is almost always subjected to high vacuum during examination, and therefore most of the samples have been either dried or impregnated with a resin. For the work illustrated here, a wide variety of techniques was used; these are fully explained in our companion volume, referred to here as *Techniques*,* which should be consulted if necessary. The stereoscopic pairs are printed at 65-mm separation, which should suit the average reader. They should be viewed with a pocket stereoscope, using spectacles if appropriate. Some people have difficulty in obtaining the three-dimensional effect; if so, they should practise on original photographs under tuition. Appendix 1 contains some notes on the terms used.

In general, the information in the captions is presented as follows: Type of sample; preparation. Discussion. Supplier of micrograph. (Where published.) Supplier of sample. Negative number, microscopist, location of microscope, sample number. (p=picture width; and θ=tilt angle.) G-series micrographs were taken by a consortium of J. M. Bell, S. R. Silva, and P. Smart; Cambridge micrographs numbered ddd.dd and Bddd.dd were taken by A. Balodis, K. Y. Wong, N. K. Tovey, and occasionally others.

During the 17 years we have been working on this topic, we have incurred many debts of gratitude, which we thankfully acknowledge. A formal list is given in the Preface of *Techniques*; but many of the names will be found in the captions here. We apologize sincerely for any omissions, which are purely accidental. Addresses of contributors are listed at the end of this volume.

* P. Smart and N. K. Tovey: *Electron microscopy of soils and sediments: Techniques*. Clarendon Press, Oxford.

1 Dispersed clay particles

When studying soil structure in the electron microscope, many of the features seen are built up from clay particles, i.e. particles smaller than 2 μm equivalent spherical diameter. The forms taken by these particles depend principally on both their mineral composition and the conditions under which they were formed. Therefore, it is usually wise to examine individual particles, prepared by dispersion, to see what they are like. The micrographs shown here and in the next Section are intended as a brief reminder of or introduction to the forms these particles may take. The principal difficulties with dispersed particles are: the particles may not have been separated; the particles may have been broken; the particles have often been 'cleaned up' by chemical treatment; and the particles may have been dehydrated or otherwise changed. For further examples, see the comprehensive collections of Beutelspacher and van der Marel (1968) and Gard (1971).

Fig. 1.1. Deflocculated Cornish kaolin; sprayed onto a mica substrate; Pt-C shadowed at 30°; stripped off; TEM; negative print. The bar in this and all subsequent micrographs is 1 μm long, unless the contrary is stated. Note the well-formed hexagonal shape of these particular crystals, much better than many kaolinites. Many of the edges are parallel, suggesting either large irregular particles or smaller ones which are intergrown with merging crystal lattices. The irregular dark marks are extinction effects resulting from either dislocations in or curvature of the crystals, and these effects may be either natural or resulting from the electron bombardment in the microscope. The black lines along the upper left-hand edges are probably fringes resulting from an electron-optical effect. Some of the other black lines nearer the right-hand side could be shadows; but the main shadowing effect has been lost in the reversal of contrast (cf. *Techniques*, Fig. 1.5). Supplied by N. A. Climpson. Negative number 9BA3, microscopist C. Golley, location English Clays Lovering Pochin and Co. Ltd.

1.2

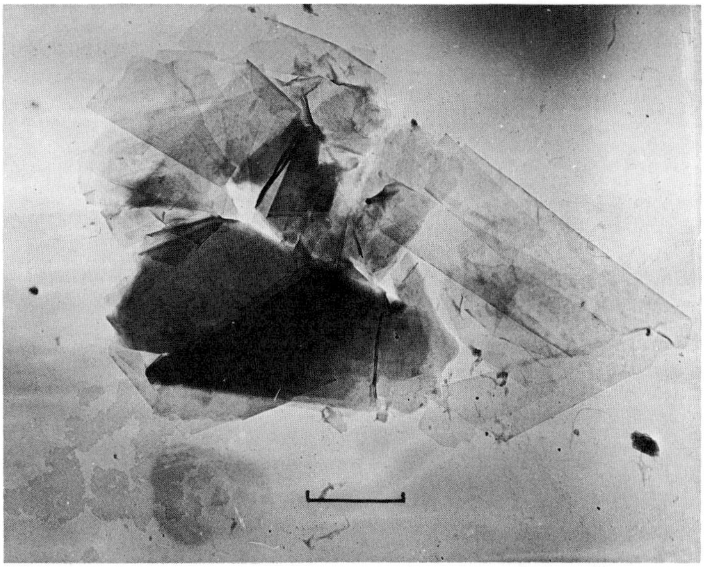

1.3

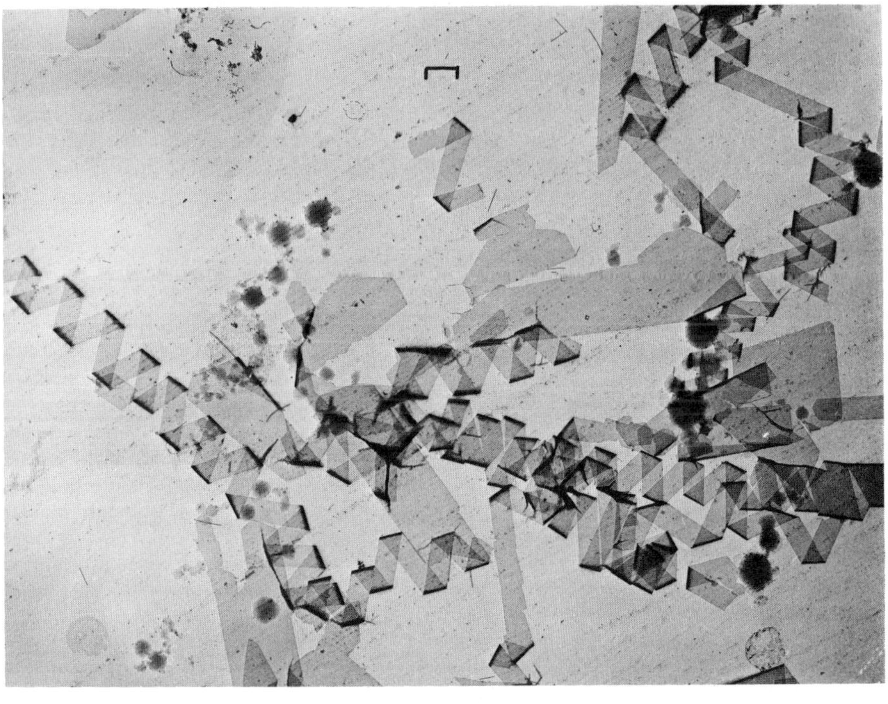

Fig. 1.2. Montmorillonite from Skyrvedalen, Norway; dispersed particles; TEM. Note the thin flake-like morphology and the folding. For technical details of the sample see Weir (1965). Supplied by A. H. Weir. Neg. 7706, R. D. Woods, Rothamsted.

Fig. 1.3. Rectorite (allevardite) from Allevard, France, also described as beidellite; dispersed particles; TEM. Note the long folded ribbons. Supplied by A. H. Weir. (From Brown and Weir 1963; see also Weir, Nixon, and Woods 1960.) 1083, R. D. Woods, Rothamsted.

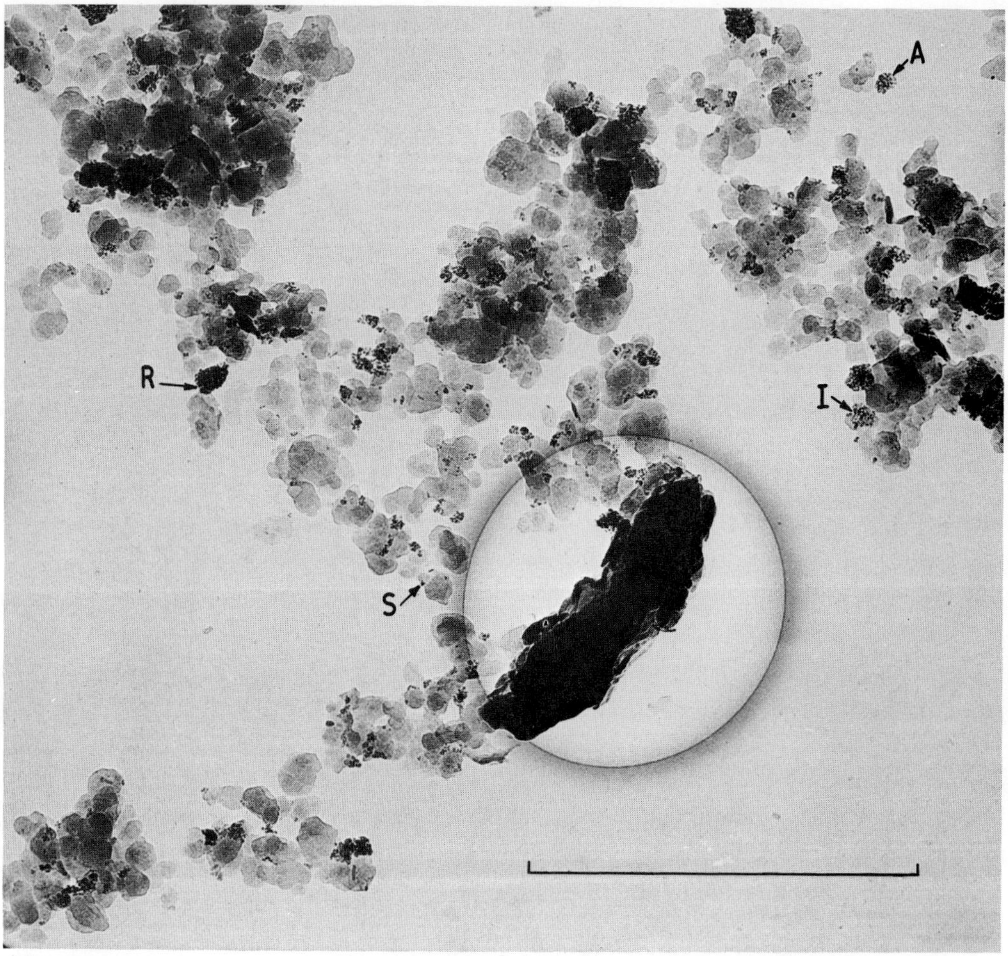

1.4

Fig. 1.4. Tropical red clay from Nyeri, Kenya; ultrasonic dispersion; TEM. This micrograph shows three types of particles: (i) one dark particle, almost 1 μm long, within the circular artefact; possibly quartz or felspar; (ii) many small plates, about 0.1 μm ϕ × 5 nm thick, somewhat variable in size and shape; almost all seen face-on, grouped face-to-face, edge-to-edge, or singly; either metahalloysite or disordered kaolinite; (iii) many very small dark particles, about 5 nm ϕ; in regular groups, R, irregular groups, I, or singly, S; usually on the small plates, sometimes alone, A; iron oxide. Sample supplied by the Road Research Laboratory. (From Smart, P. (1972). Nyeri Red Clay. Institution of Civil Engineers dissertation (technical essay).) 1821, P.S., Glasgow, sample 288.5.

Fig. 1.5. Chrysotile asbestos fibres or tubes from Jeffrey Mine, Canada; micropulverized; embedded in methyl methacrylate, via ethanol and ethanol–methacrylate mixtures; ultrathin section cut using a diamond knife in an ultramicrotome; mounted on carbon film; methacrylate dissolved away with benzene; TEM. (a) shows a group of fibres cut almost perpendicular to their axes. Fringes are visible where the electron beam is closely parallel to the lattice planes; but oxygen and hydrogen ions may have been removed by electron bombardment from the structureless outer region, e.g. near point A. Otherwise, these tubes may be on the point of coalescing. Fibre F shows an unusual growth pattern. (b) shows details of tubes A and B. Note the multispiral circumferential fringes and the zig-zag radial ones. A few dislocations, in which the circumferential fringes split in two, may just be discernible in the right-hand tube. Supplied by K. Yada. (From Yada 1967.) Tohoku University. ▷

1.5a
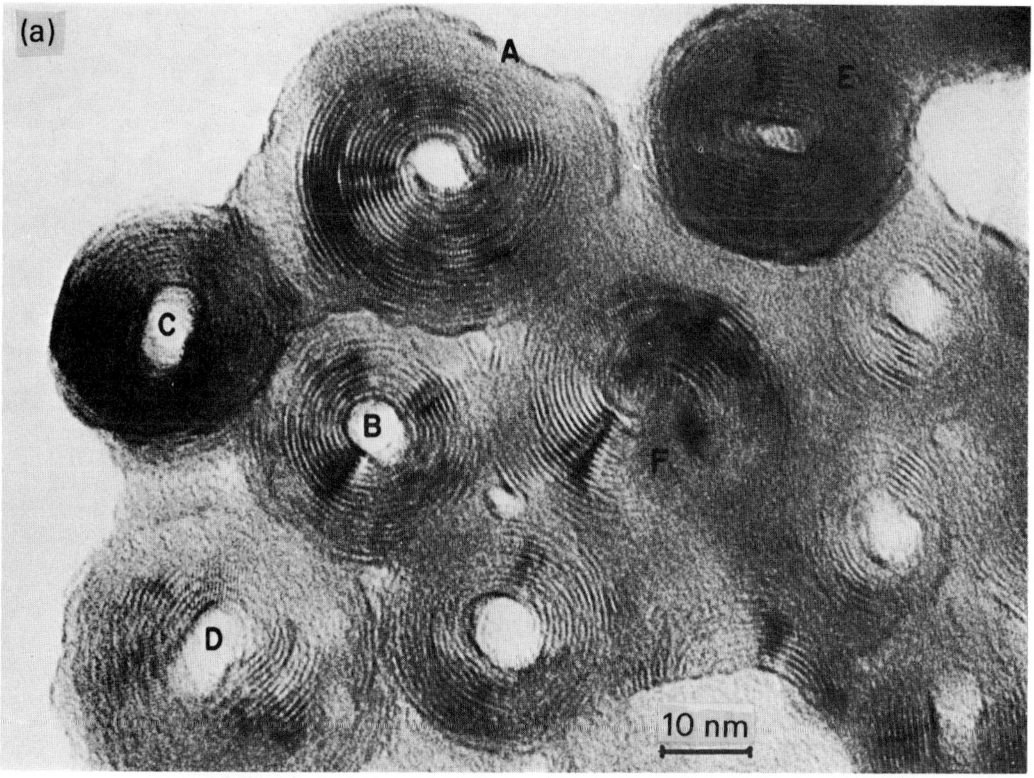

1.5b
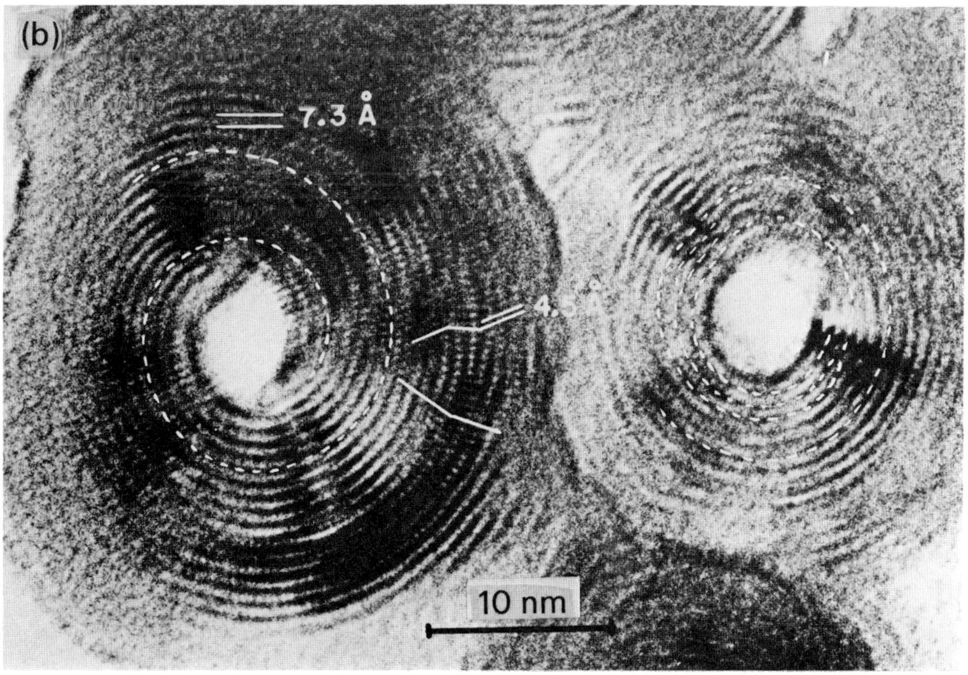

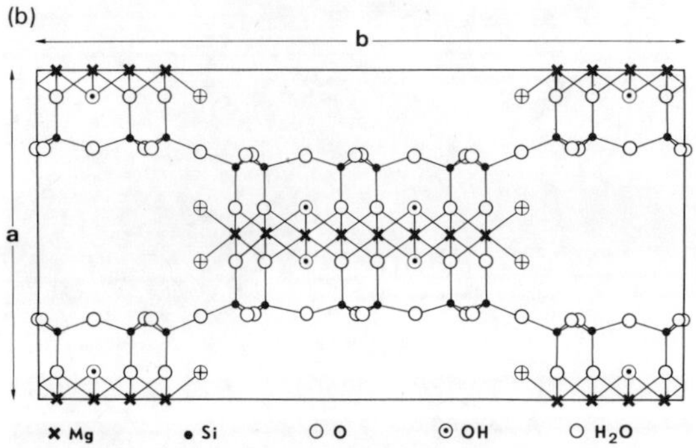

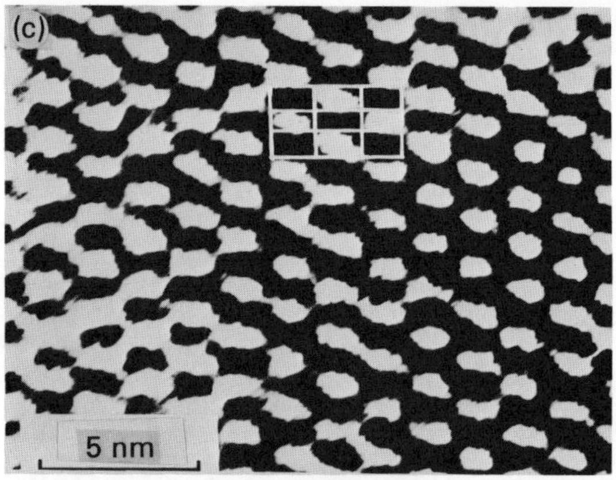

Fig. 1.6. Sepiolite fibres from Ampandrandava, Madagascar; impregnated using vapour phase transfer; ultrathin section, 55 nm thick, perpendicular to the fibre axis; TEMs using 'lattice imaging', viz. (a) interference between the 000, $0\overset{\pm}{2}0$, and $1\overset{\pm}{1}0$ beams. (c) between 000, $\overset{\pm}{0}20$, $\overset{\pm\pm}{1}10$, $\overset{\pm}{2}00$, and $\overset{\pm\pm}{1}30$. Note the quincunx-like structure, i.e. there are four units at the corners of a rectangle and a fifth at the centre, see the simplified ionic structure in (b). From X-ray analysis, $a \approx 1.3$ nm and $b \approx 2.7$ nm, cf. (c). Electron diffraction suggested refinements to the structure, which, together with the internal porosity, gave the sepiolite a zeolitic character. Supplied by M. Rautureau. (From Rautureau and Tchoubar 1976; for lattice imaging see Allpress and Sanders 1973.)

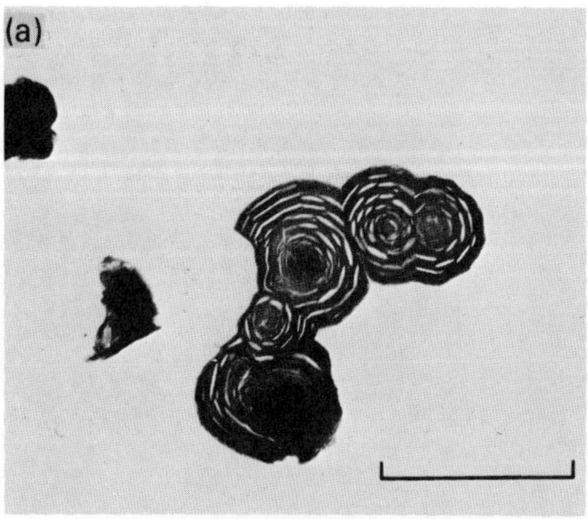

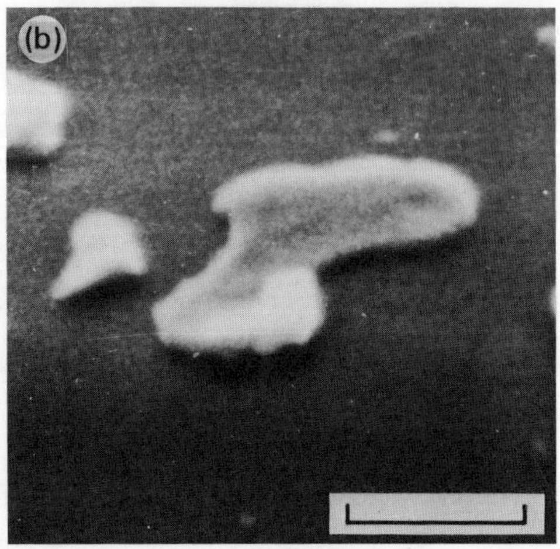

Fig. 1.7. Halloysite from Pahoia Tuff, North Island, New Zealand; ultrasonic dispersion in water with NH_4OH at pH 10; particles > 1 μm centrifuged away; spotted onto carbon coated formvar films on copper grids. (a) TEM. (b) The same particles: coated with gold; SEM. The halloysite forms discs, which are sandwiched between silica flakes. Supplied by J. H. Kirkman. (From Kirkman 1977.) A. S. Craig, DSIR, Palmerston North.

2 Undisturbed clay minerals

To supplement the previous Section, a small collection of micrographs showing 'typical' clay particles *in situ* is included. Note, however, that many of these have been chosen because they show well-formed crystals. Fig. 2.14 is included here for convenience.

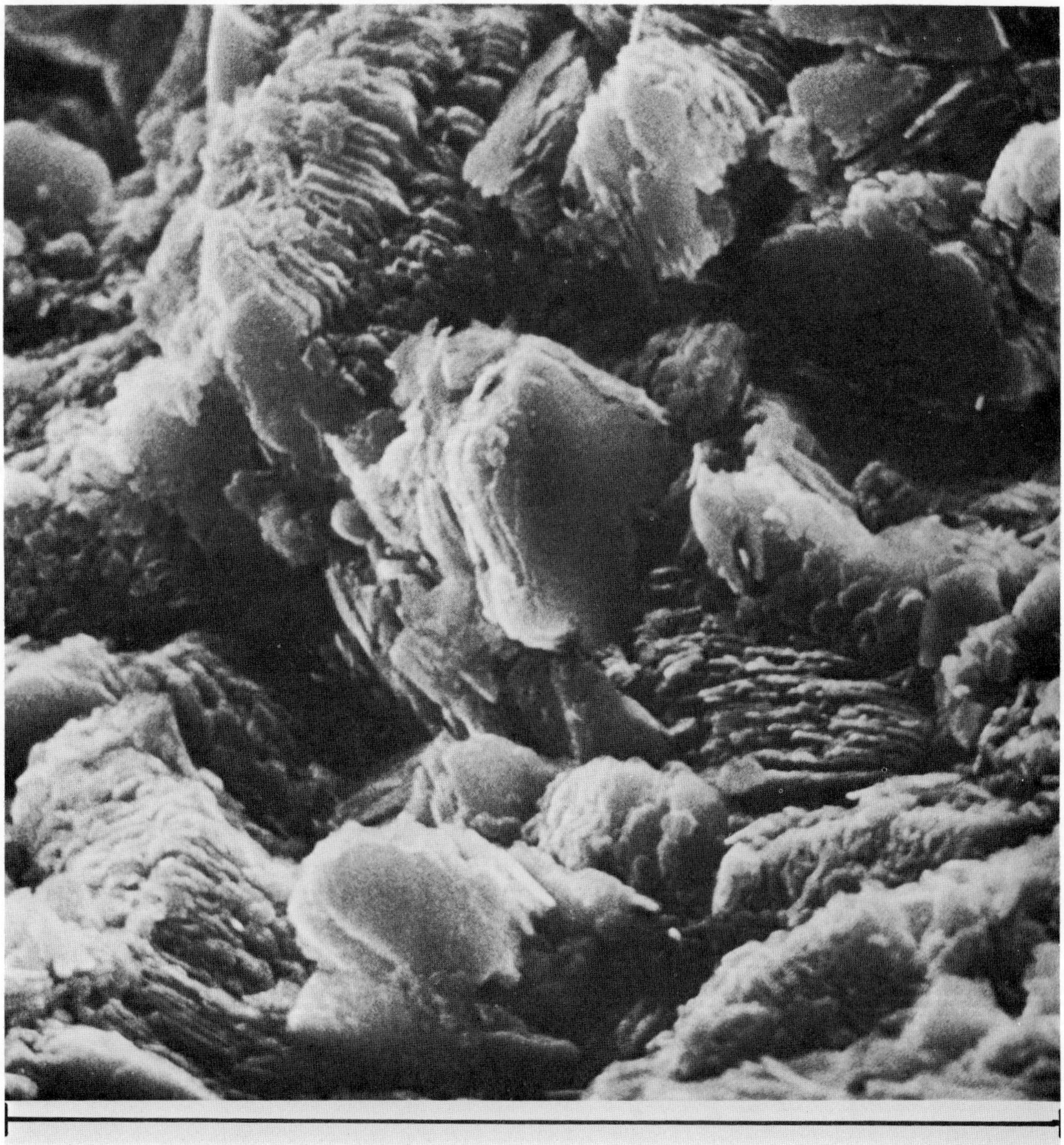

Fig. 2.1. Kaolin from Blackpool Quarry, Cornwall; undisturbed lump; air-dried; fractured; coated with 25 nm Au multidirectional, to prevent charging; SEM, θ 60°, i.e. the apparent line of sight is at 60° to the mean normal to the sample. The apparent line of illumination was inclined almost as much on the diametrically opposite side of the normal; and the secondary emission mode was used, so the illumination was rather diffuse. p 50 μm. Note the random distribution of packets of loose plates or stacks of plates firmly joined together. Sample supplied by F. R. Noble. (From Tovey, N. K. (1971). A selection of scanning electron micrographs of clays. Cambridge University Engineering Department Technical Report CUED/C—SOILS/TR5.) 82.33, Cambridge.

Fig. 2.2. As Fig. 2.1; except stereoscopic pair of micrographs, p 17 µm, θ left 40°, θ right 45°. For stereoscopic viewing (with most microscopes), the micrographs must be turned through 90° from their position for monoscopic viewing. (From Tovey, N. K. (1971). A selection of scanning electron micrographs of clays. Cambridge University Engineering Department Technical Report CUED/C—SOILS/TR5.) 289.19/20, Cambridge.

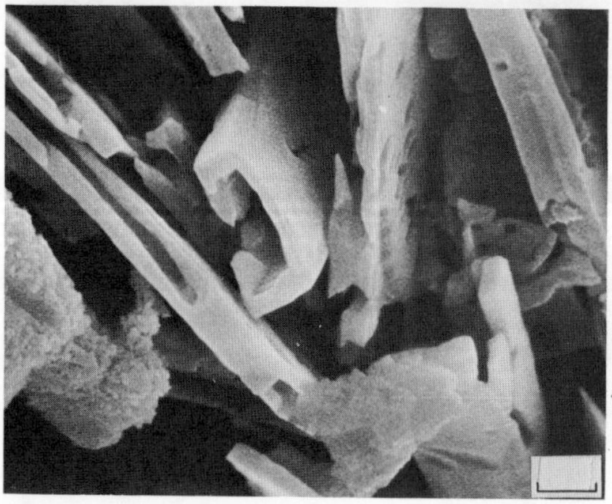

Fig. 2.3. Non-consolidated natural kaolinite from Mesa Alta, New Mexico; ion-beam thinned; TEM. The micrograph shows the gentleness of ion-beam thinning, which is suitable for extremely brittle samples such as this, for which the central assemblage or stack is only connected to the matrix by very few point or line contacts. Supplied by M. Paulus. (From Paulus, Dubon, and Etienne 1975.) PE 1226, Wards No. 9.

◁ **Fig. 2.4.** Halloysite from San-in District, Japan; fracture surface; coated with carbon and gold; SEM. The square-tube shape is characteristic of Tazaki's Ht3–2 class of halloysite. Supplied by K. Tazaki. (From Tazaki, 1978*b*.)

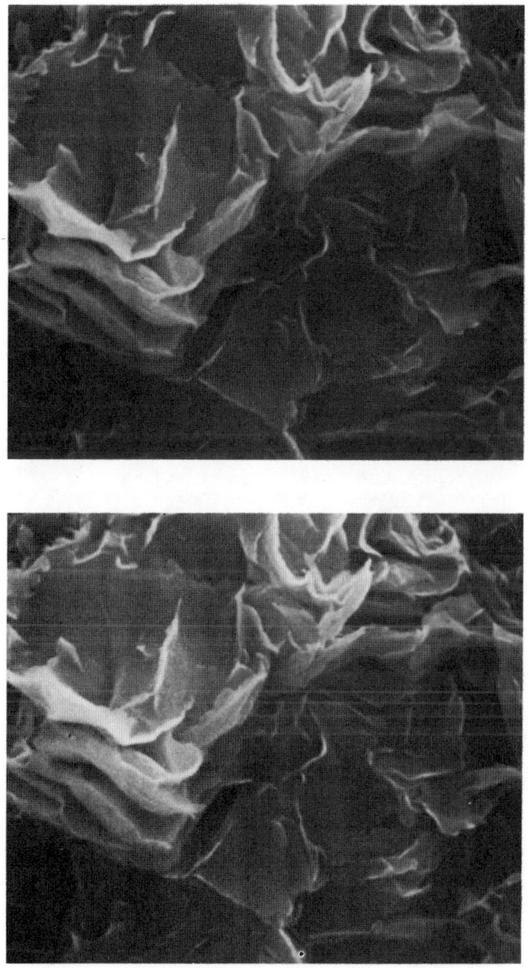

Fig. 2.5. Montmorillonite (bentonite) from Clay Spur, Wyoming; air-dried; fractured; 25 nm Au multi-directional; stereoscopic SEMs, p 6.8 µm, θ 50/55°. The wafers are either individual particles or quasi-crystals; note their characteristic tendency to curl. (From Tovey, N. K. (1971). A selection of scanning electron micrographs of clays. Cambridge University Engineering Department Technical Report CUED/C—SOILS/TR5.) 324.4/5 Cambridge.

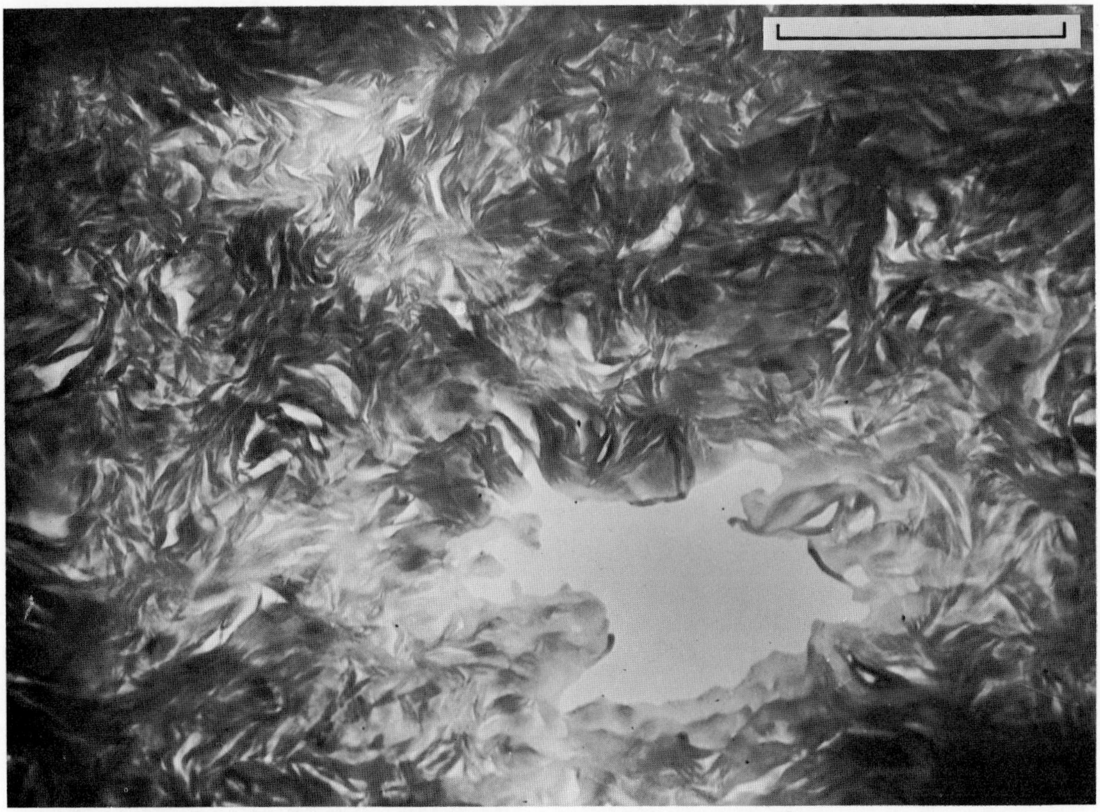

Fig. 2.6. Non-consolidated natural montmorillonite from Amory, Mississippi; ion-beam thinned; TEM. This micrograph is included to demonstrate that ion-beam thinning can be used for montmorillonite. The process is usually allowed to continue until a hole about 1 μm ϕ appears in the sample. Note the contorted nature and close packing of the particles. Supplied by M. Paulus. (From Paulus, Dubon, and Etienne 1975.) PE 1280, Wards No. 22A.

Fig. 2.7. Illite from Morris, Illinois; air-dried; fractured; 25 nm Au multidirectional; SEM, p 7.5 μm, θ 60°. Note the three-dimensional appearance of many of the particles. (From Tovey, N. K. (1971). A selection of scanning electron micrographs of clays. Cambridge University Engineering Department Technical Report CUED/C—SOILS/TR5.) 335.8, Cambridge.

Fig. 2.8. As Fig. 2.7, θ 60°, but p 15 μm. Some particles are planar, thinner than kaolinite plates, and with less tendency to curl than montmorillonite. (From Tovey, N. K. (1971). A selection of scanning electron micrographs of clays. Cambridge University Engineering Department Technical Report CUED/C—SOILS/TR5.) 329.41, Cambridge.

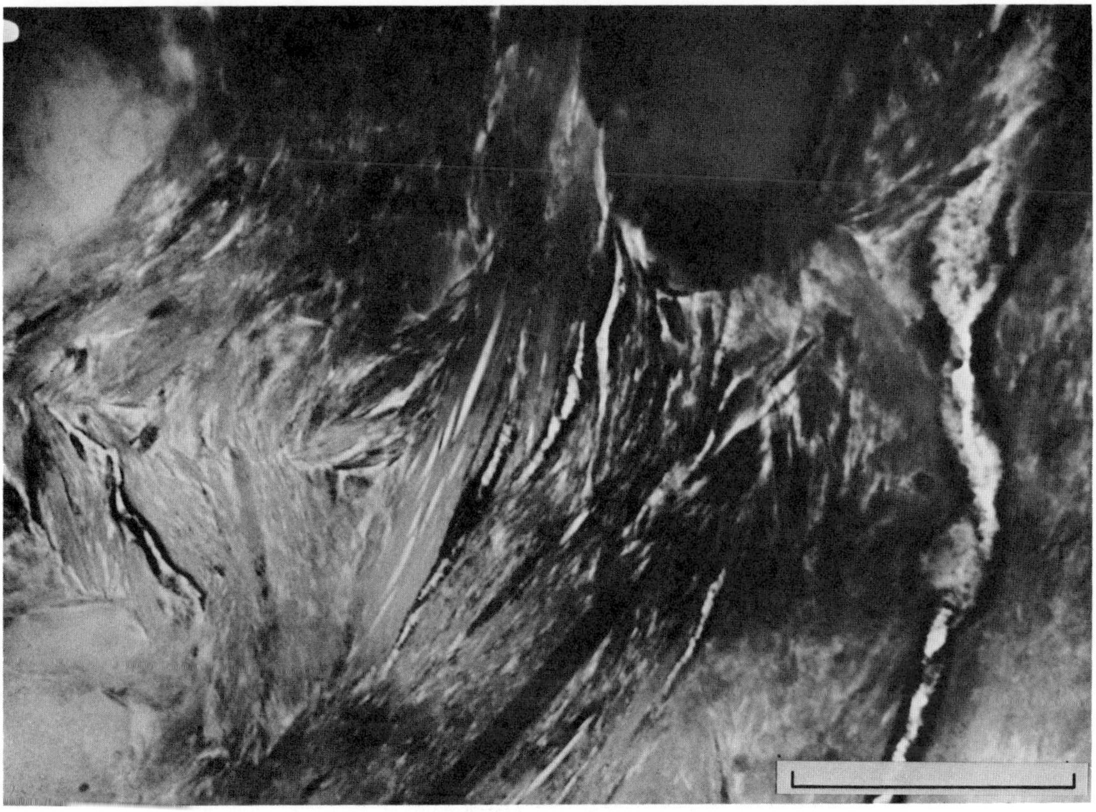

Fig. 2.9. Non-consolidated natural illite from Fithian, Illinois; ion-beam thinned; TEM. Many of the clay plates are seen edge-on and almost parallel to each other in domains. Supplied by M. Paulus. (From Paulus, Dubon, and Etienne 1975.) PE 1232, Wards No. 35.

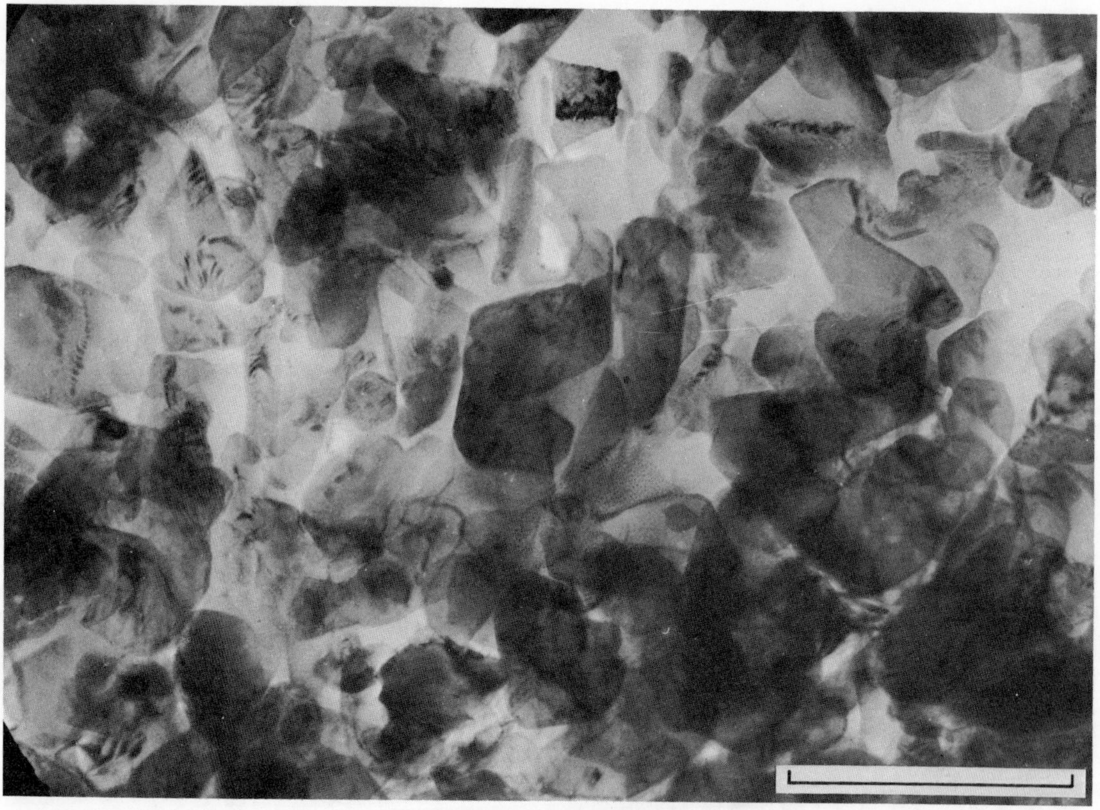

Fig. 2.10. Non-consolidated natural chlorite; ion-beam thinned; TEM. The plate-like particles are seen here face-on, almost parallel to each other, and forming a domain; this interpretation was confirmed in the original work by another section at right angles. Supplied by M. Paulus. (From Paulus, Dubon, and Etienne 1975.) PE 1305, Guelta Boring 104.

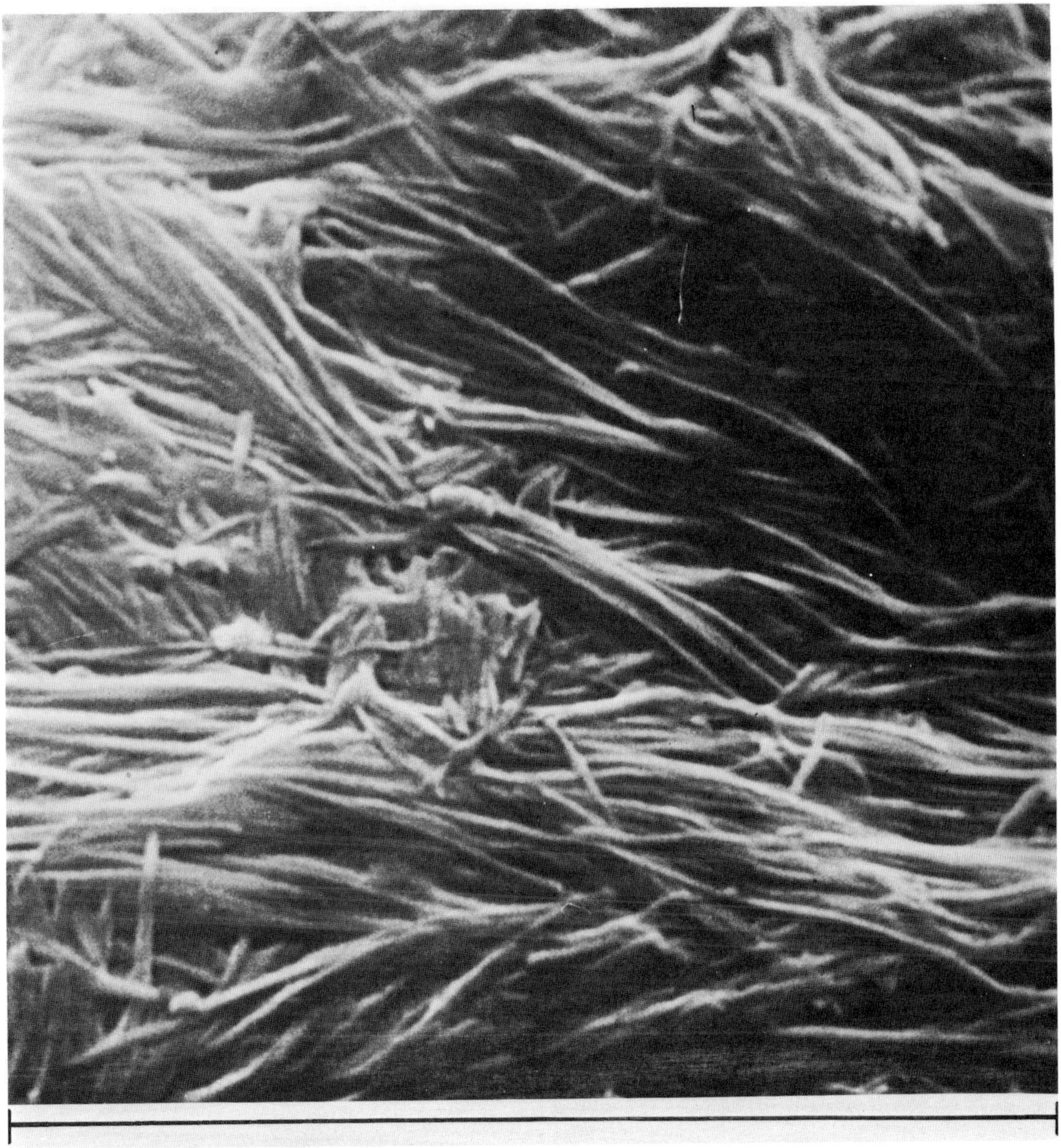

Fig. 2.11. Attapulgite from Attapulgus, Georgia, USA; undisturbed; air-dried; fracture surface; 25 nm Au multidirectional; SEM, p 4.7 μm, θ 60°. The needle-like particles form interwoven sub-parallel bundles. (From Tovey, N. K. (1970). Electron microscopy of clays. Ph.D. thesis, University of Cambridge.) 326.16, Cambridge.

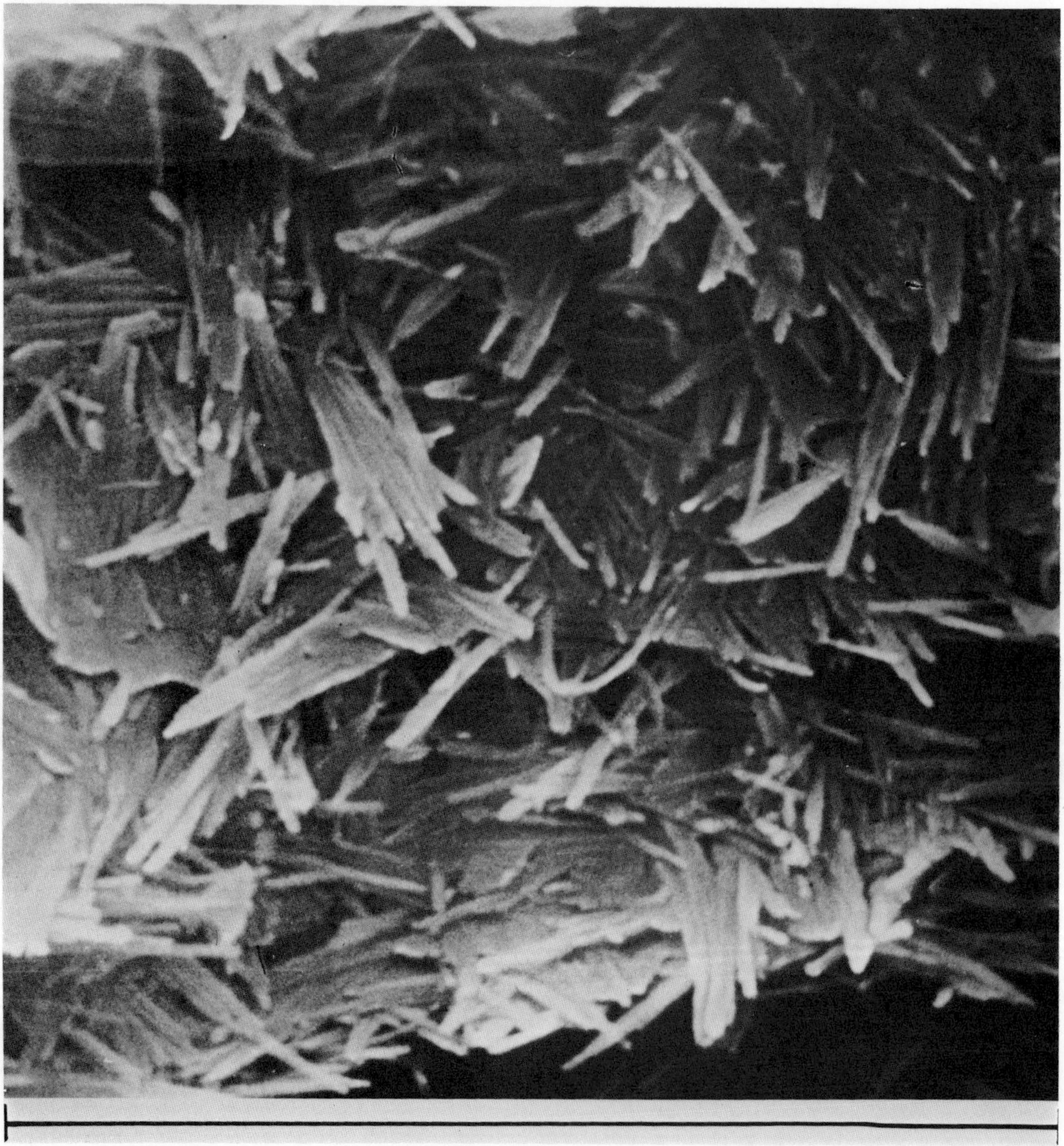

Fig. 2.12. Same as Fig. 2.11. This shows a tendency for the needles to aggregate into plates. 334.27.

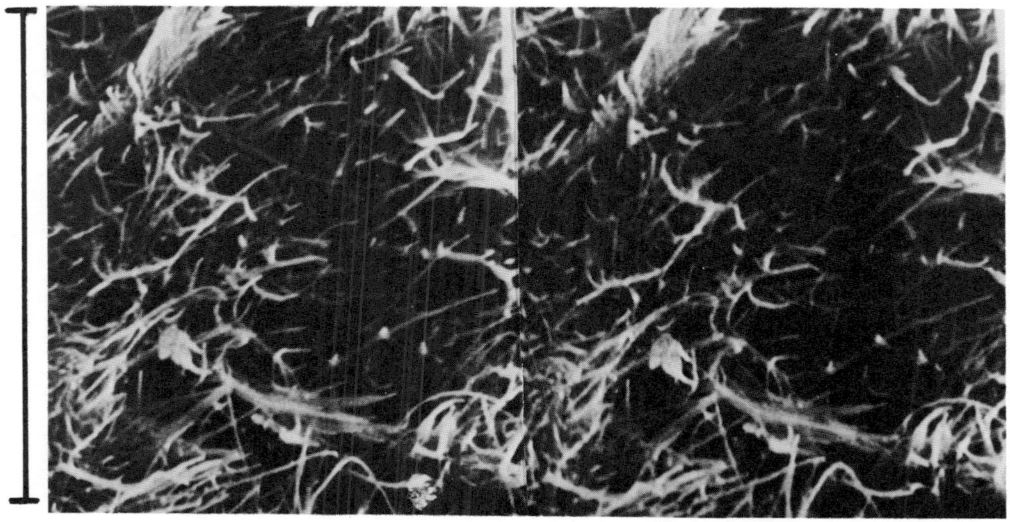

Fig. 2.13. Sepiolite from Madrid, Spain; undisturbed; air-dried; fracture surface; 20 nm Au multidirectional; SEM, *p* 7.5 μm, θ 33/38°. The fibres occur individually and are interwoven irregularly to form a mat. (From Tovey, N. K. (1971). A selection of scanning electron micrographs of clays. Cambridge University Engineering Department Technical Report CUED/C—SOILS/TR5.) 421.39 and 421.40, Cambridge.

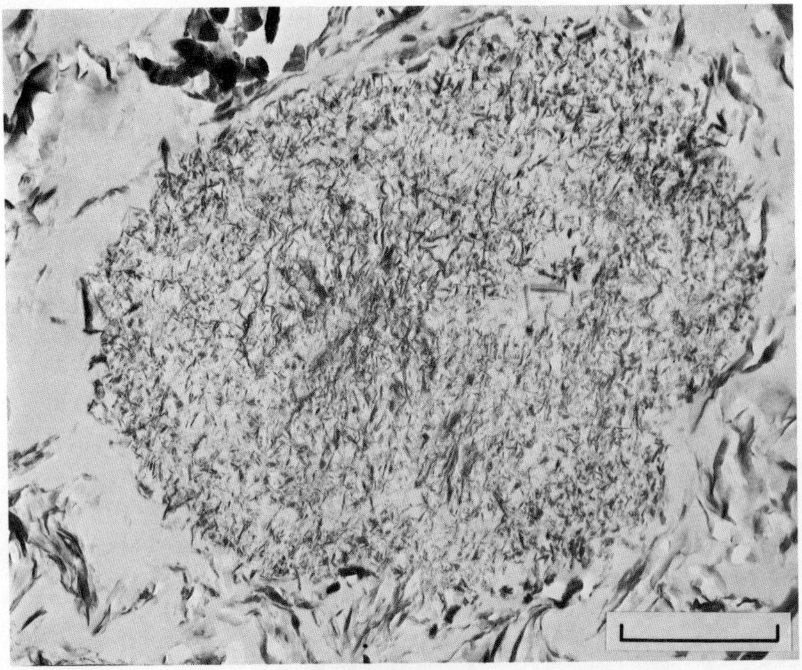

Fig. 2.14. Zeolitic aggregate in a yellow–brown late-Cretaceous clay from the Pacific floor, 11° 14.7′ N, 150° 17.5′ W, 144 m depth below sea floor under 5330 m water; remoulded sample; pore water substituted by ethyl alcohol, amyl acetate, liquid CO_2; supercritically dried; vacuum impregnated with Spurr resin; ultrathin section; TEM by R. H. Bennett of NOAA. (From Bennett, Bryant, and Keller 1977.) Deep Sea Drilling Project Leg 16 Site 163A.

3 Laboratory studies of sedimentation

One of the earliest hypotheses concerning the structure of clay sediments was, in its simplest form, as follows: the clay particles are plates with negative electrostatic charges bound on their faces and positive charges bound on their edges; the positive edges link onto the negative faces; physicochemical or mechanical restraints stabilize these contacts in such a way that the plates meet at an angle of, say, 60°; so that a single-plate card-house structure is formed. Later, it was realized that, under certain conditions, both the edges and the faces of the plates become negatively charged; and, it was deduced that, under these conditions, the particles are piled up face-to-face, this arrangement being sometimes termed dispersed structure. This Section illustrates some of the laboratory experiments which have been made to investigate these hypotheses. The principal experimental difficulty is to ensure and to verify that the plates have not been pushed around as the samples are prepared for observation; thus, Fig. 3.5 is an important control for freeze-drying, whilst Fig. 3.7 illustrates an alternative approach based on impregnation and sectioning.

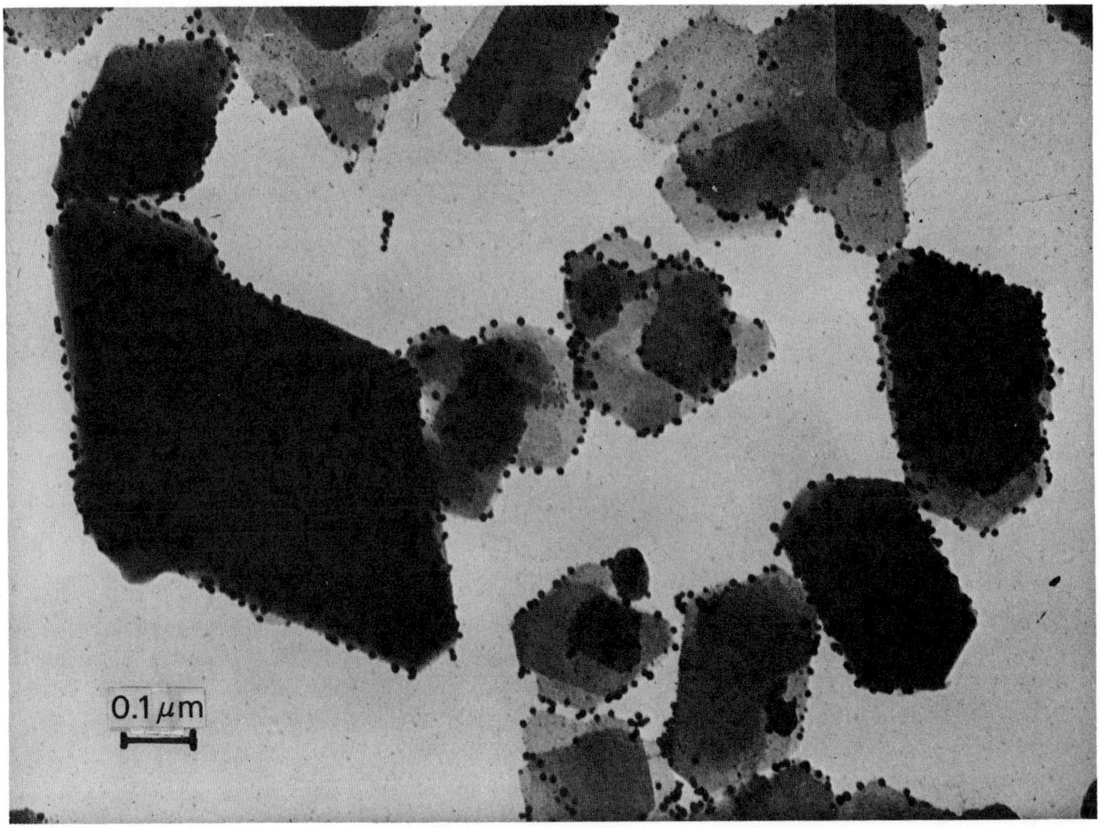

Fig. 3.1. Kaolinite, decorated with gold particles by mixing in water following Thiessen (1942); TEM. The gold particles, which are negatively charged, adhere only to the edges of the kaolinite, thus showing these edges to be positively charged, at least under the conditions of the experiment. Supplied by H. van Olphen. (From van Olphen 1977.) H. P. Studer, Shell Development Co.

Figs. 3.2 and 3.3. Hirschau kaoliniote; Fig. 3.2 dispersed in 0.01 N ammonia solution; Fig. 3.3 suspended but coagulated in 1 per cent $CaCl_2$ solution; fished out on copper oxide whiskers; dried at 50 °C; gold coated; TEM. Supplied by B. Mattiat. (From Mattiat 1969.) R. Blaschke & Frl. Vogt. Münster. ▷

3.2

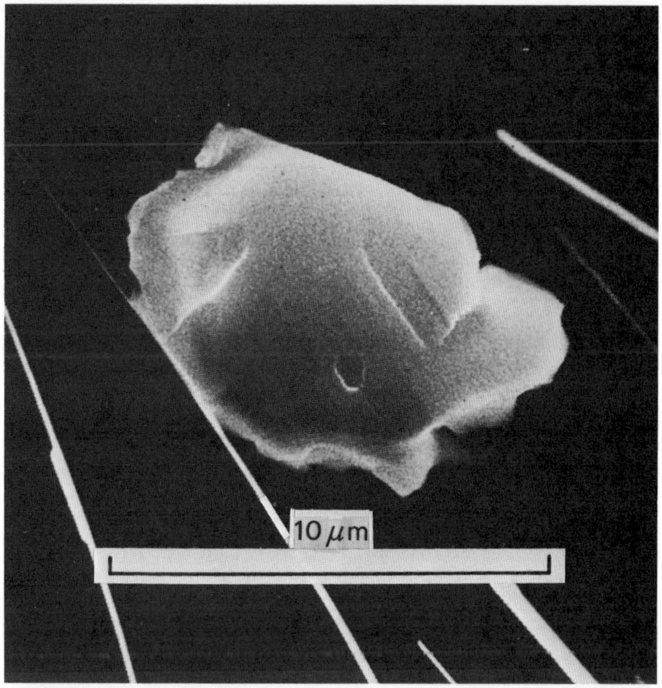

3.3

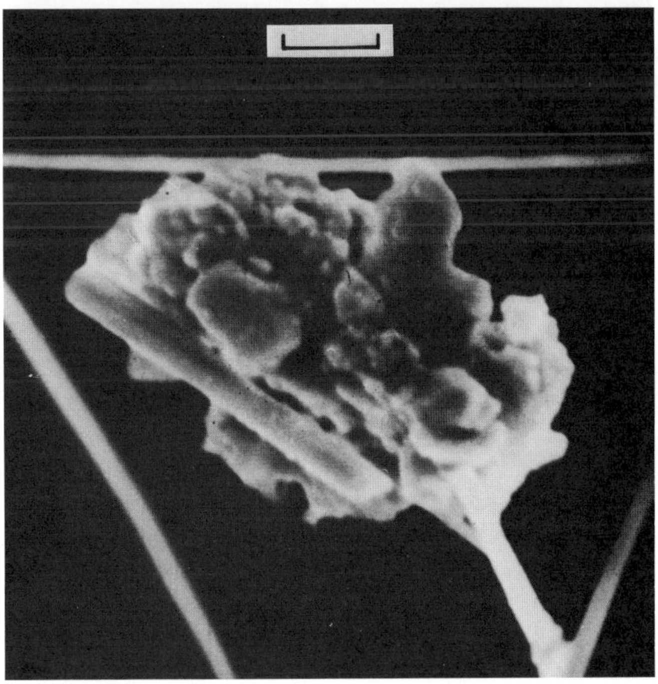

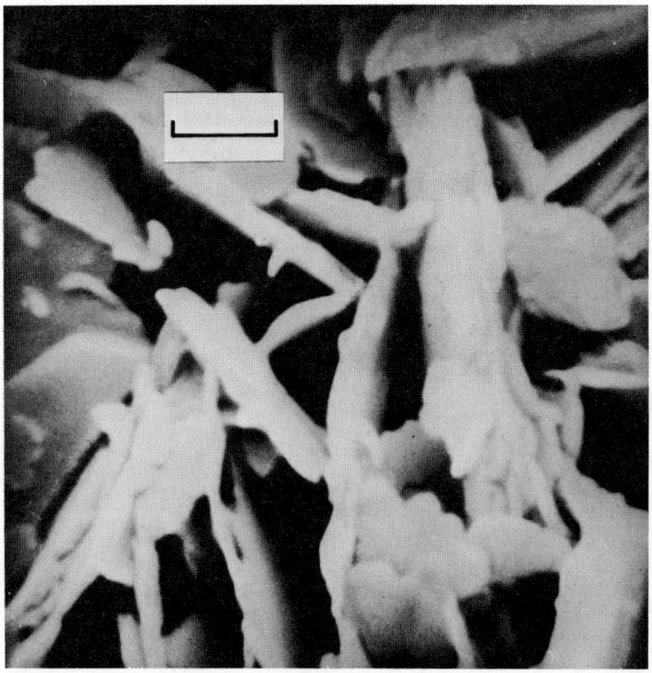

Fig. 3.4. Hirschau kaolinite, < 2 μm ϕ; sedimented in very hard water from the R. Weser; frozen in liquid propane (or a similar fluid) at 113 K or 83 K; fractured whilst frozen; dried below 193 K, 0.0001 torr; gold coated; SEM. Almost single-plate card-house structure. Supplied by B. Mattiat. (From Mattiat 1969.) R. Blaschke & Frl. Vogt, Münster.

Fig. 3.5. Geisenheim illite, < 2 μm ϕ; sedimented in 0.01 N ammonia solution; otherwise as Fig. 3.4. Dispersed structure. ▷

Fig. 3.6. Geisenheim illite, < 2 μm ϕ; sedimented in 1 per cent $CaCl_2$ solution; otherwise as Fig. 3.4. Sponge-like honeycomb structure.

3.5

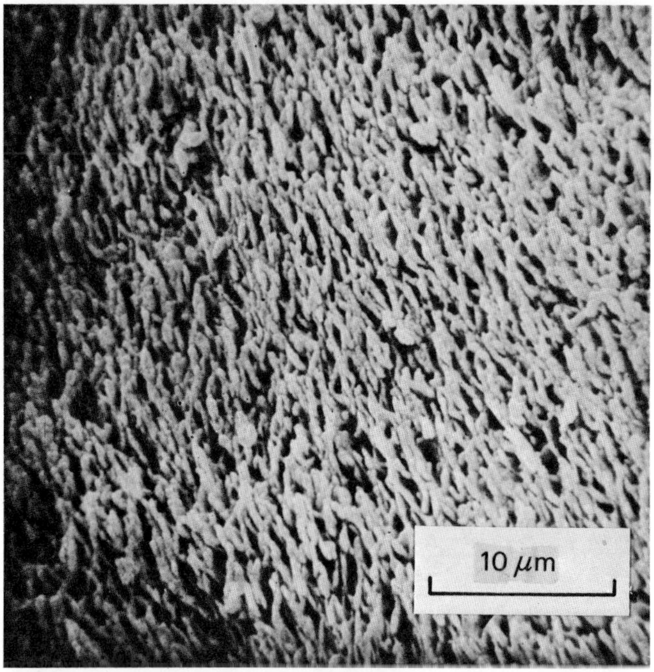

3.6

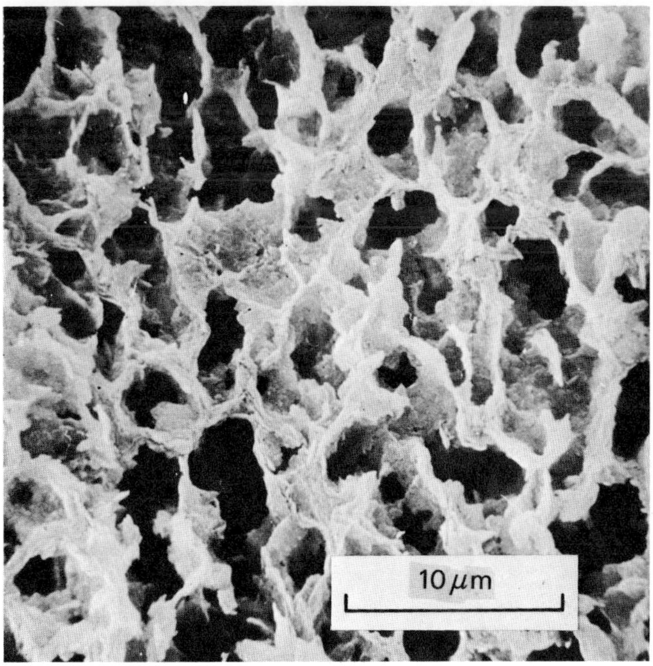

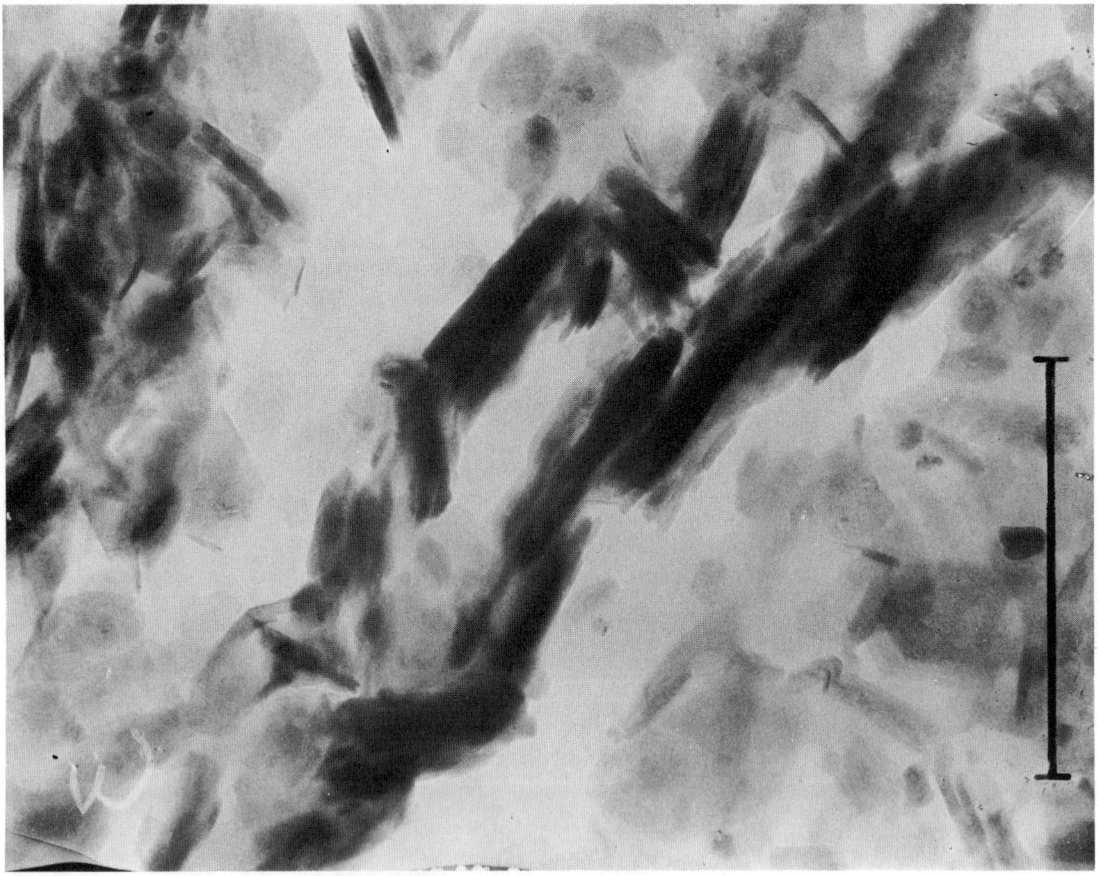

Fig. 3.7. Hydrite PX Kaolinite from Georgia; dispersed with a trace of NH_4OH at 20 g l^{-1} clay concentration flocculated at 10 g l^{-1} clay concentration and 1g l^{-1} Na Cl concentration; impregnated with polyethylene glycol; very thin section, 1 μm thick, cut with a microtome; high voltage, 500 kV, TEM. A stepped cluster of face-to-face flocculated flakes. Supplied by N. R. O'Brien. (From O'Brien 1971.) Kyoto University.

4 Inter-particle structures

The inter-particle structures of natural clay sediments depend on: the mineral type, size, and shape, of the particles; whether these are *ab initio* aggregated or fully dispersed; the concentration of particles in the water; the concentration of salts in the water; turbulence; biological reworking after sedimentation; and possibly other factors. In consequence, a variety of structures have been formed and recognized.† This Section illustrates these. Figs. 4.4–4.6 and 4.7–4.8 compare the effect of salinity.

†For the nine most important, see Appendix 1 from random to turbostratic.

4.1

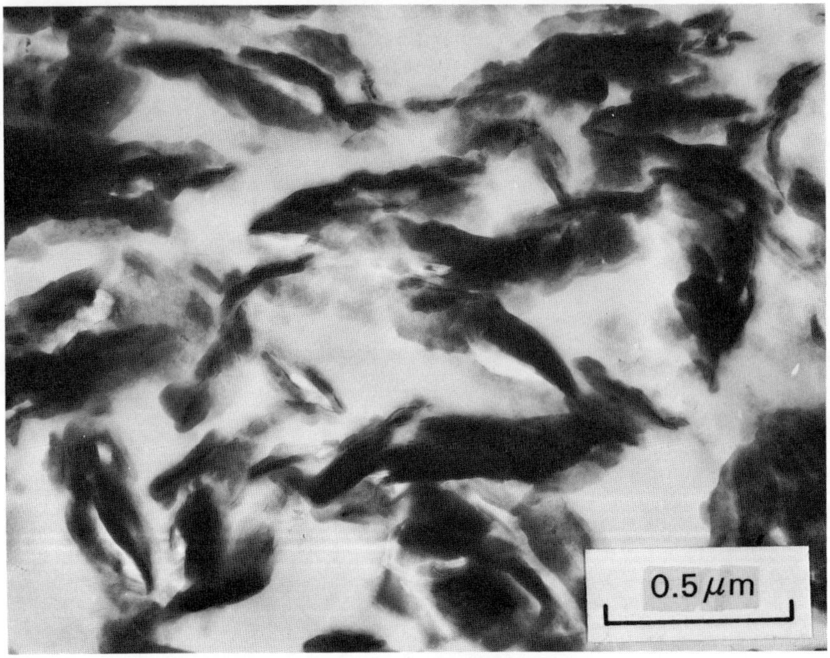

4.2

Fig. 4.1. Pacific floor clay, same sample and preparation as Fig. 2.14, except undisturbed sample. Note the remarkable open structure with edge-to-edge and edge-to-face contacts, presumably controlled by both mineral type and particle size. On the left, a zeolitic aggregate; upper right-hand corner, a white circular artefact 1 μm ϕ. Technical data: 81 per cent clay, 18 per cent silt, clay dominantly smectite, void ratio 3.5. (From Bennett *et al.* 1977.) R. H. Bennett of NOAA, Texas A & M Univ., DSDP.16.163A.

Fig. 4.2. Submarine sediment from Mississippi Birdfoot Prodelta from 27.7 m depth below sea floor under 70 m water; undisturbed sample; preparation as Fig. 2.14. This micrograph shows almost, but not quite, single-plate card-house structure, which tends towards book-house structure or discrete domain structure. (From Bennett *et al.* 1977.) R. H. Bennett of NOAA, Texas A & M Univ., McClelland Boring B-1.

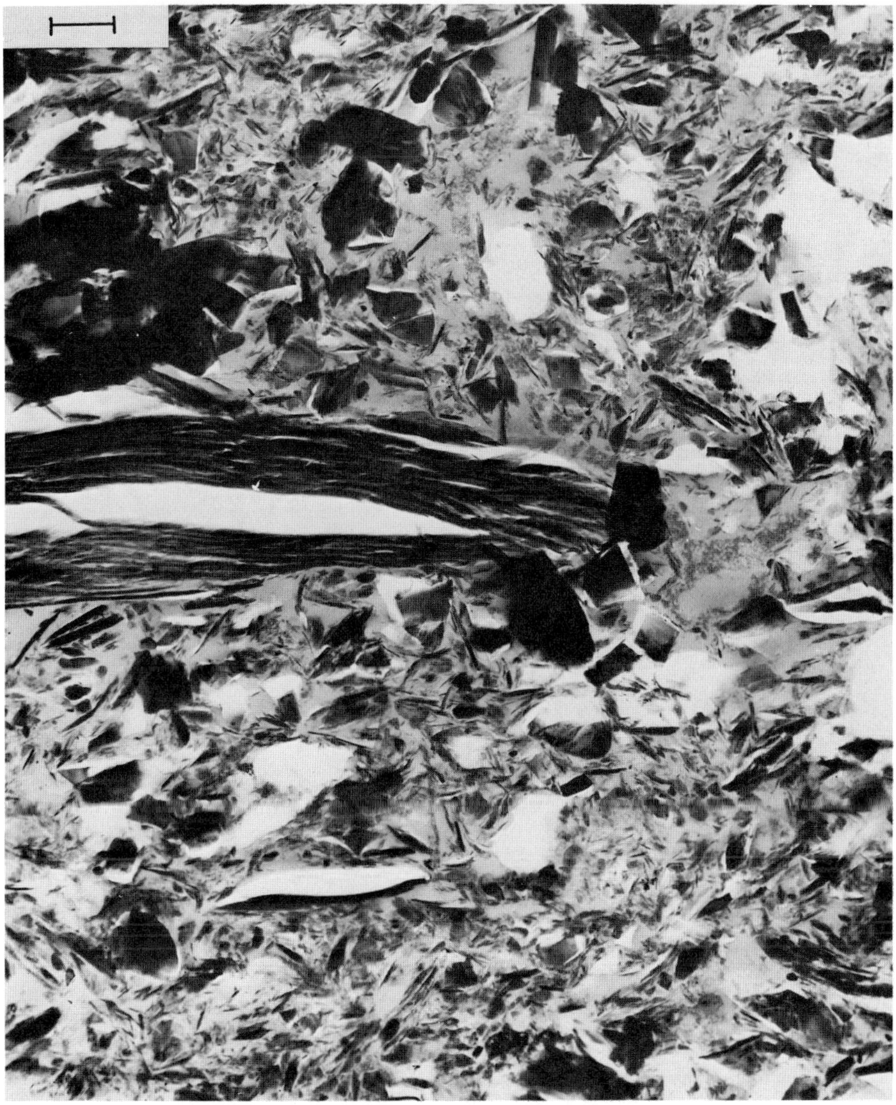

Fig. 4.3. Leda Clay, Ottawa Sewage Works, depth 15 m (50 ft); undisturbed; immersed in alcohol; impregnated with Araldite AY18; ultrathin section, vertical section, vertically cut; TEM. The large particle running across the centre of the micrograph has split during microtomy; also, at its right-hand end, there is a dark particle 1 μm ϕ, which overlies other details, and which is therefore probably a loose particle, i.e. an artefact. From this particle and curving up to the left of a white hole in the section 1 μm ϕ, is a band of poorly defined rectangles say 0.25 μm wide by 0.4 μm; these are thought to be the remains of a microfossil. To the right and below the loose particle is an elliptical void roughly 1 μm across surrounded by amorphous material; this feature is thought to be of biological origin. In general, note the random arrangement of the smaller particles; with particles seen both edge-on and face-on; and with oblique edge-to-face contacts between the particles: but note also that these smaller particles are in pairs or small groups face-to-face. The structure, therefore, was classified as salt-flocculated or book-house rather than single-plate card-house. (From Smart, P. (1966). Soil structure, mechanical properties and electronmicroscopy. Ph.D. thesis, University of Cambridge.) Sample obtained through C. B. Crawford. G.2995, Cambridge, 94.101.

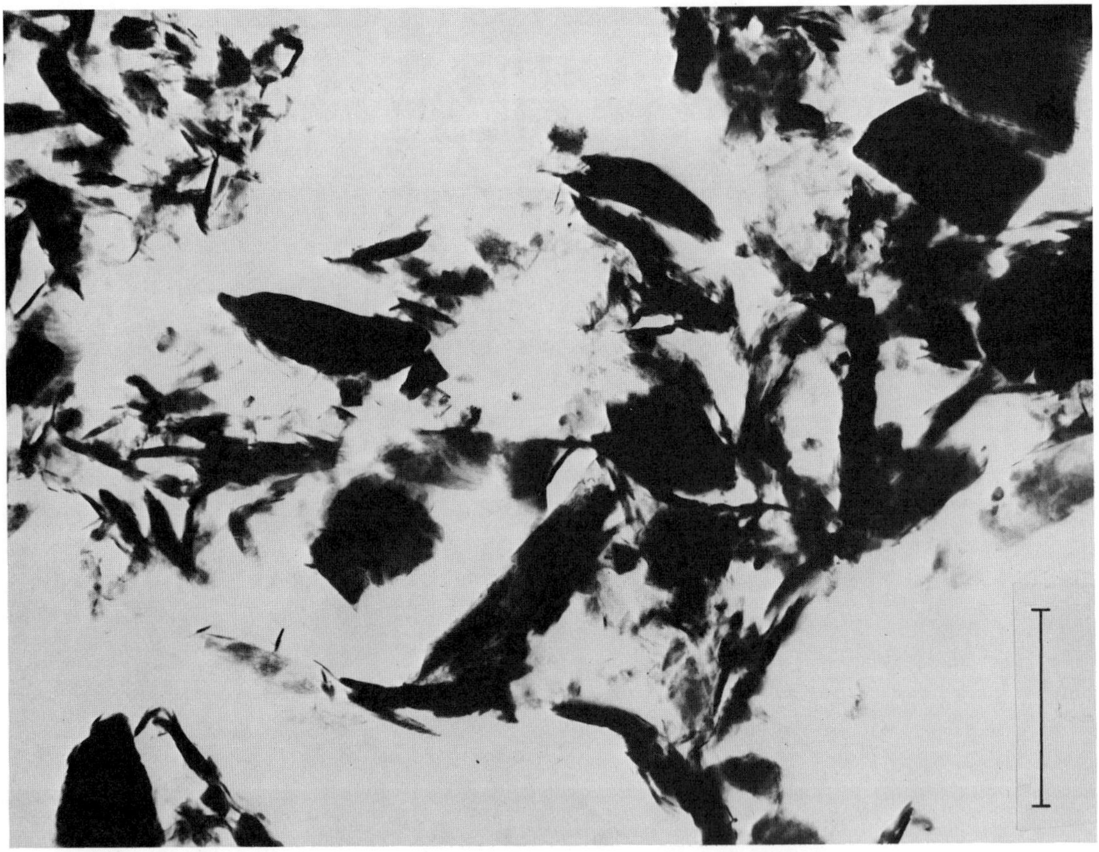

Fig. 4.4. Quick marine clay from Lilla Edet, 6 m depth; undisturbed sample; impregnated; ultrathin section; TEM. An example of flocculent structure. Note that the particles are grouped in flocs, which are themselves grouped loosely. Compare this marine clay with the next two examples. Approximate technical data: $\rho=1.5$ tm^{-3}; w = 91 per cent; LL = 58 per cent; PL = 29 per cent; S_t = 147; natural groundwater level = 1 m below ground level. Micrograph supplied by R. Pusch.

Fig. 4.5. Brackish-water clay from Skå-Edeby, 2 m depth; undisturbed sample; acrylate impregnated; ultrathin section; TEM. An example of 'Pusch's' structure, but cf. Fig. 4.9. Note: (i) aggregates of various size and density and rigidity; (ii) links, i.e. small, fairly loose particle groups connecting the larger, stiffer aggregates; and (iii) the large variation in pore sizes. Micrograph supplied by R. Pusch. ▷

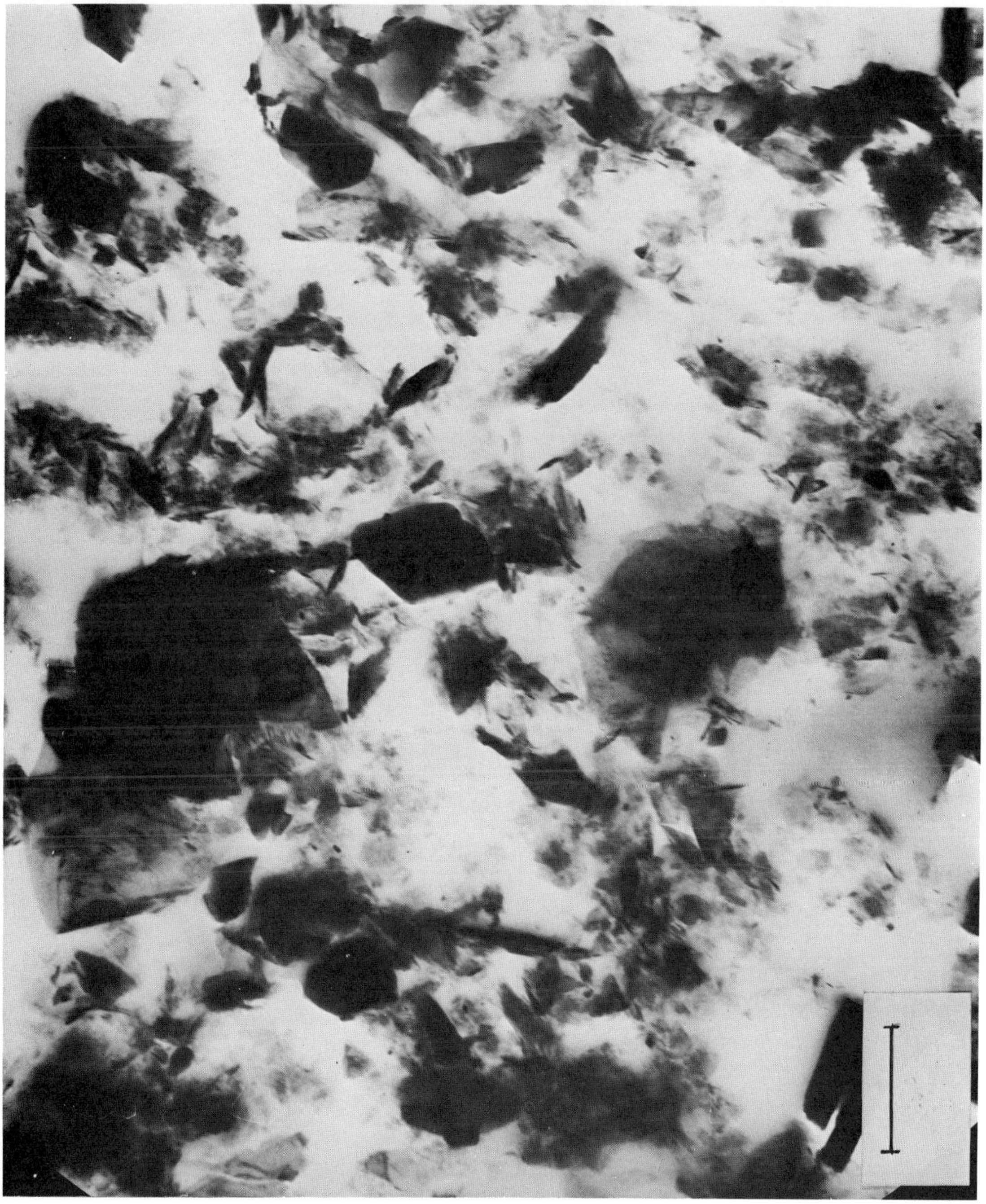

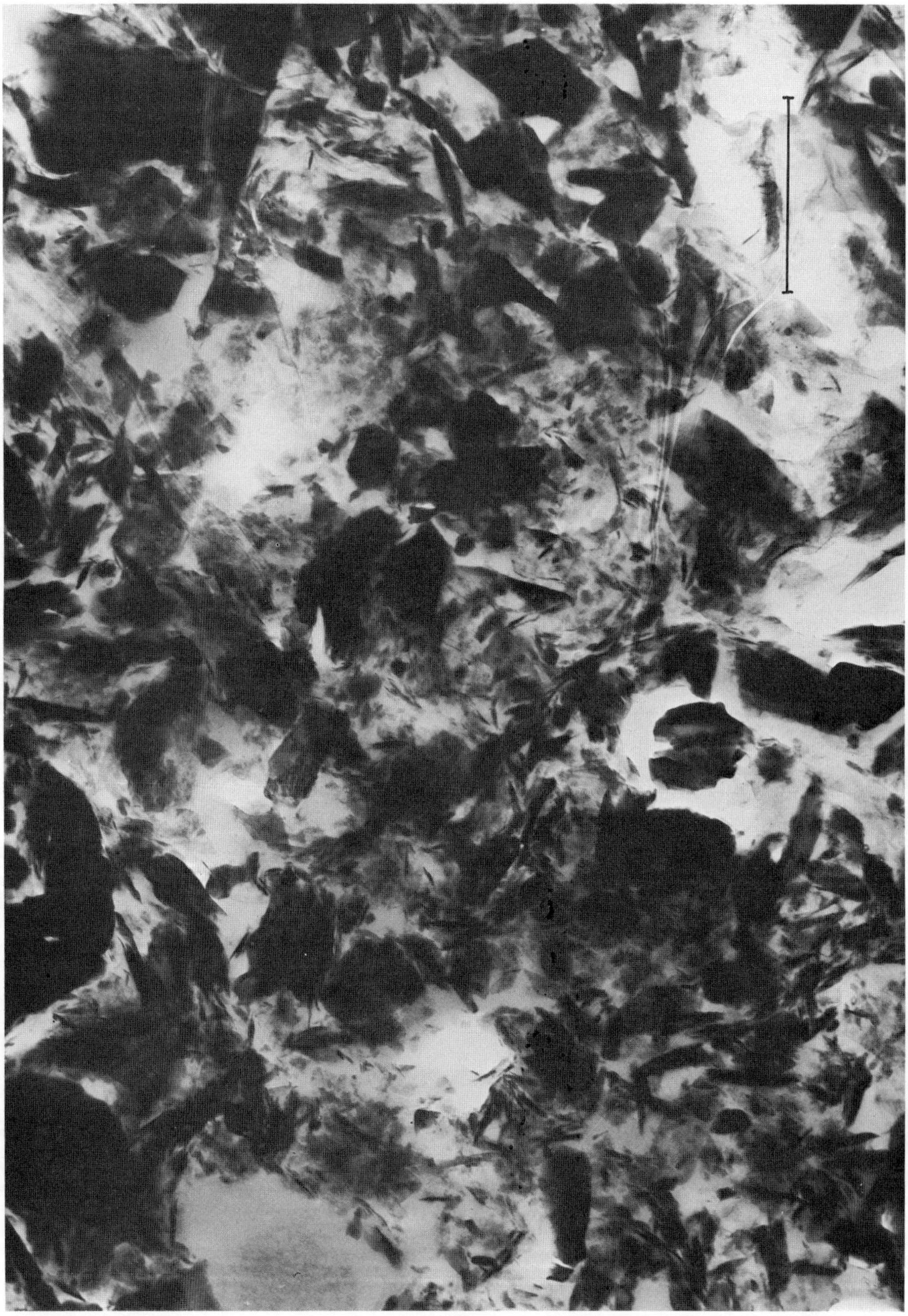

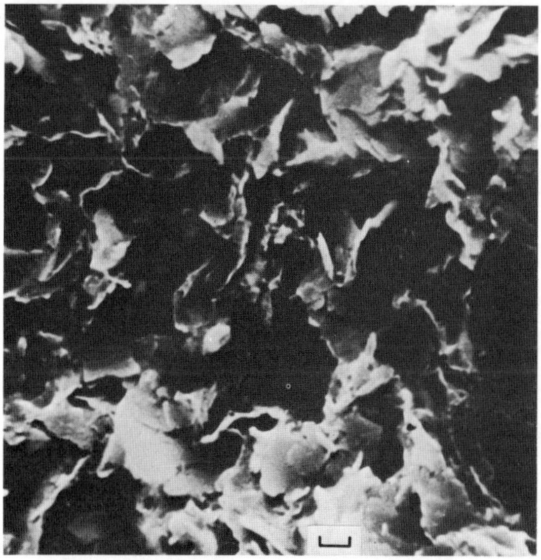

Fig. 4.7. North Sea mud from the Jadebusen; undisturbed sample; freeze-dried as Fig. 3.4; SEM. Almost single-plate card-house structure. Supplied by B. Mattiat. (From Mattiat 1969.) R. Blaschke & Frl. Vogt, Münster.

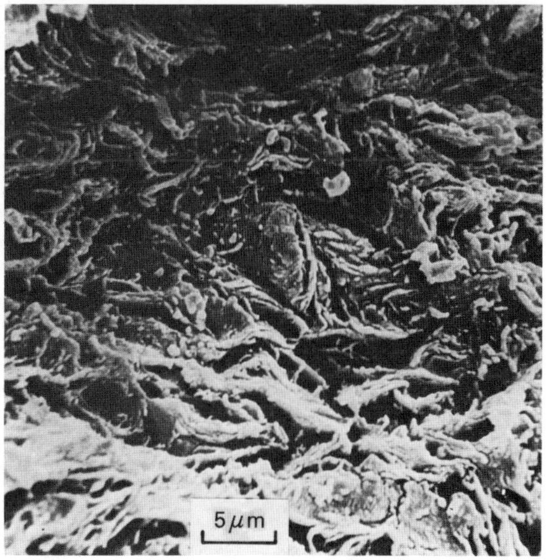

Fig. 4.8. Brack-marsh sediment from Hatzum; otherwise as Fig. 4.7. Dispersed or turbostratic structure.

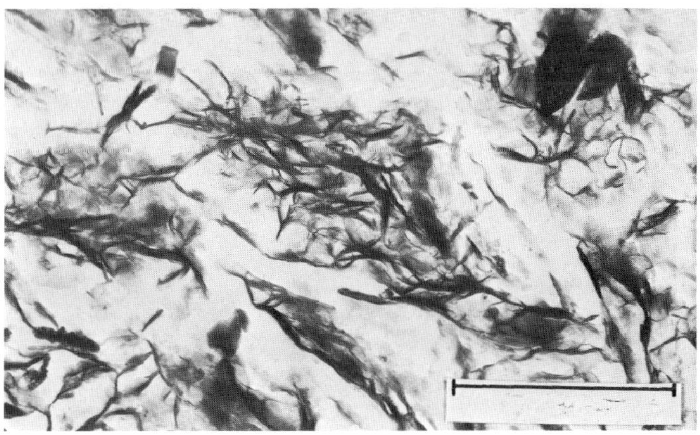

Fig. 4.9. Pacific floor clay, same sample and preparation as Fig. 2.14, except undisturbed sample. Note the flocculent structure with denser micro-aggregates within it, Pusch's structure. (From Bennett *et al.* 1977.) R. H. Bennett of NOAA, Texas A & M Univ., DSDP.16.163A.

◁ **Fig. 4.6.** Freshwater clay from Skå-Edeby, 10 m depth; undisturbed sample; impregnated; ultrathin section; TEM. An example of almost random structure. Streaks, such as those near the micrometre mark are photographic defects. Technical data: $\rho = 1.6$ tm^{-3}; $w = 65$ per cent; LL = 52 per cent; PL = 22 per cent; $\tau_{fn} = 10$ kPa; $S_t = 16$; groundwater level = 0.5 mbgl. Micrograph supplied by R. Pusch.

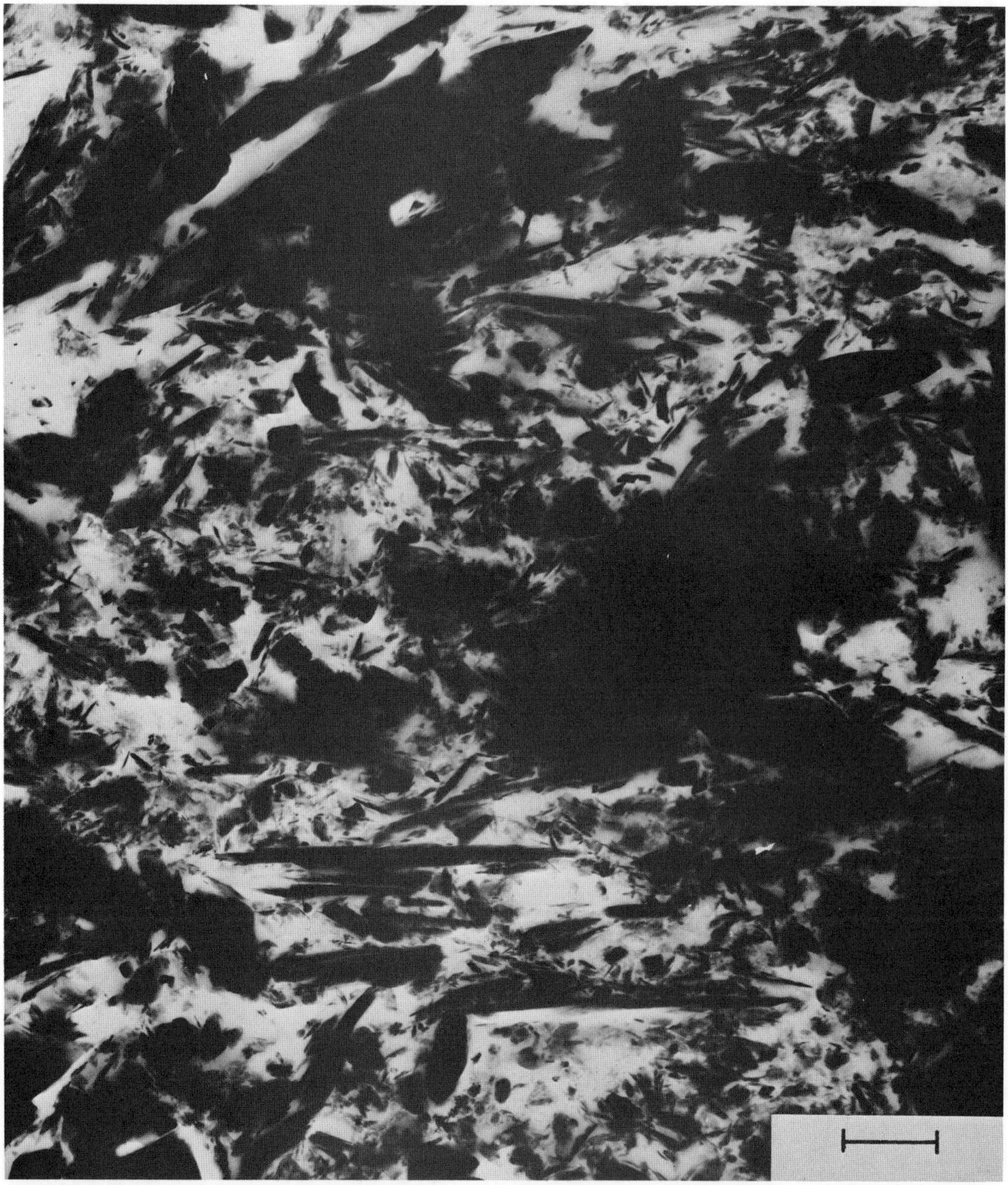

Fig. 4.10. Leda Clay, Ottawa Sewage Works, depth 15 m (50 ft); undisturbed sample; immersed in water; substituted with methanol; impregnated with Araldite AY18; ultrathin section, vertical section, vertical cut; TEM. Whilst observing, it was thought that the larger particles, which are seen edge-on in the lower part of the micrograph, are horizontal, but it was not possible to decide definitely. Accepting this interpretation, the micrograph shows a tendency for the larger plate-like particles to be horizontal or sub-horizontal, whilst the smaller particles are arranged almost at random. This is in accordance with the definition of Brownian structure. (From Smart, P. (1966). Soil structure, mechanical properties and electronmicroscopy. Ph.D. thesis, University of Cambridge.) Sample obtained through C. B. Crawford. G.4011, Cambridge, 94.101.

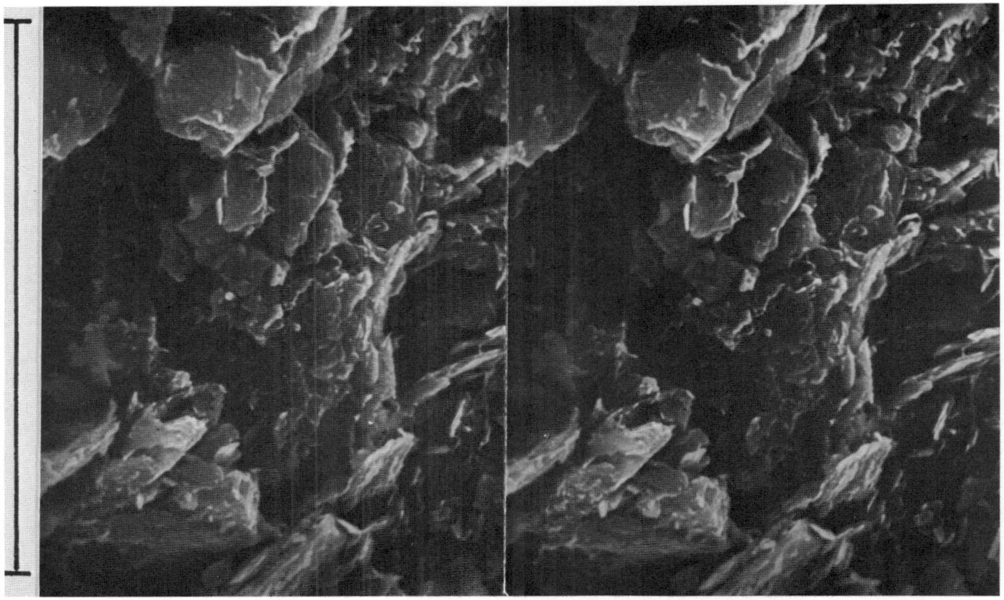

Fig. 4.11. Drammen Clay from Konnerudgt, Norway; undisturbed; air-dried; fractured; 10 nm C, 20 nm Au; stereoscopic SEMs, p 20 μm, θ 40/45°. This was selected as an example of a so-called clay arrangement; the structure is almost single-plate card-house with some plates face-to-face and a little fine silt. 240.40 and 240.39, Cambridge.

Fig. 4.12. Undisturbed 'clay'; air-dried; vertical fracture; 25 nm Au; SEM, p 20 μm, θ 30°. This is a granular arrangement of equant silt particles. The grooves on the particle to the left of centre are probably evidence of glaciation. Sample supplied by A. N. Schofield. 313.26, Cambridge.

5 Consolidation

At one time there was considerable interest in the effect of preferred particle orientation on the anisotropy of the mechanical properties of clays. A number of laboratory studies, therefore, set out to investigate the development of preferred particle orientation during consolidation; the term is used here with its engineering sense, i.e. the gradual reduction of volume of a saturated sediment as the water is squeezed out by a pressure, usually vertical. Some of these studies are illustrated here. We should perhaps remark that it was later remembered that in many cases in nature, larger scale features, some of which can be seen by eye, are more important in their effect on mechanical anisotropy; and the impetus was lost. However, some interesting phenomena remain to be explained; and some of the quantitative methods of analysis mentioned in *Techniques*, Chapter 12 would be applicable to a study of these.

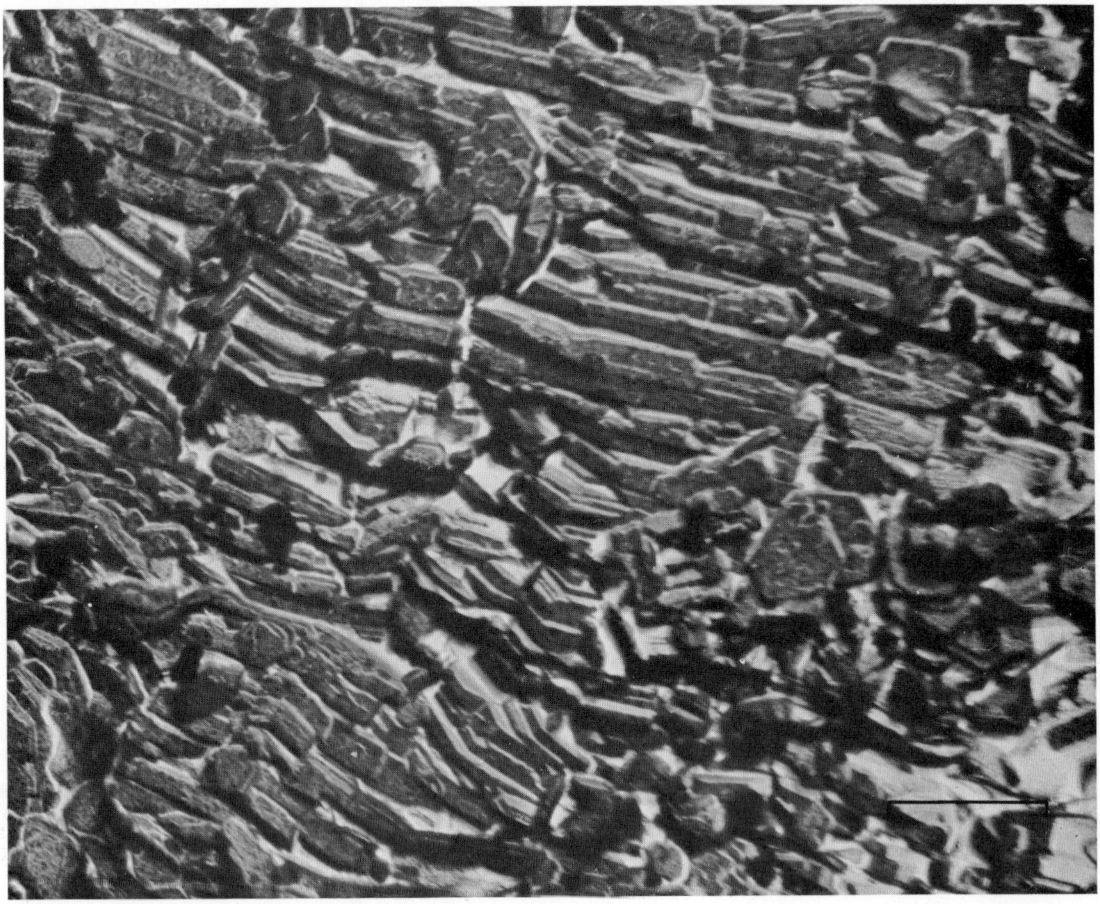

Fig. 5.1. Georgia kaolin; ground; suspended in water; made homoionic to H^+ and OH^- using ion-exchange resins; 0.33 per cent aluminium sulphate added as a flocculant; filter pressed at 172 kPa (25 p.s.i.) for 30 min; dried; replica of the edge of the filter cake; TEM. Note the orientation of the particles, which are mainly seen edge-on; replicas of the top surface of the cake showed the particles face-on. Deflocculated clay had similar orientation with slightly closer packing. Supplied by R. West. (From West 1963.)

Figs. 5.2–5.4. Spestone Kaolin; remoulded at 160 per cent mc; consolidated uniaxially to 0, 200, 500 kPa respectively; vertical surfaces; SEMs, *p* 16 μm; θ 50, $\approx$ 0, 45° respectively; Fig. 5.3 the vertical is up-down as printed; Fig. 5.4 horizontal view, i.e. the apparent line of sight is tilted in the horizontal plane and the vertical is across. Note the progressive development of preferred horizontal orientation, which was confirmed by stereoscopy for Fig. 5.3, and by looking along the preferred direction in Fig. 5.4. Details: freeze-dried, air-dried, substitution-dried with ether respectively; Figs. 5.2 and 5.3. fractured, 10 nm C bidirectional, 20nm Au multidirectional; Fig. 5.4. peeled, 20 nm Au multidirectional; these differences are thought to be inconsequential. (From Tovey, N. K. (1971). A selection of scanning electron micrographs of clays. Cambridge University Engineering Department Technical Report CUED/C—SOILS/TR5.) 213.6, 205.28, and 163.24, Cambridge.

5.3

5.4

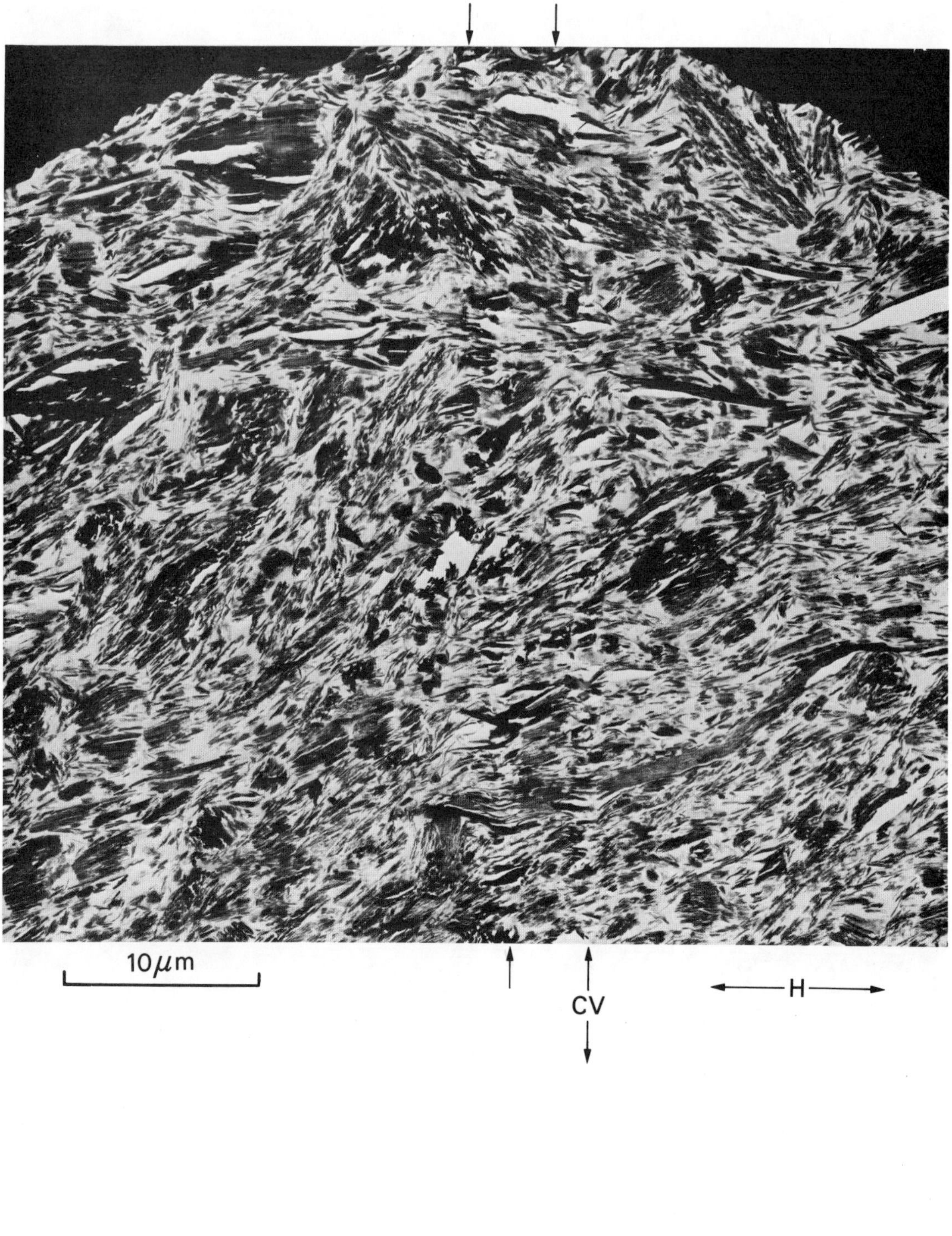
10μm
CV
H

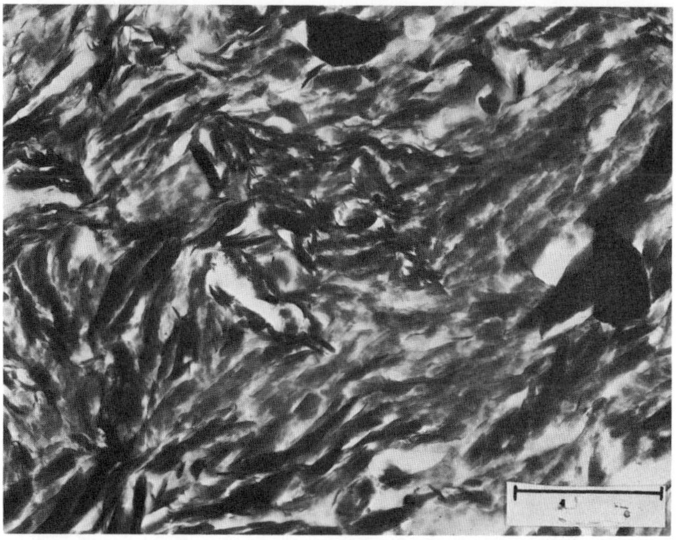

◁ **Fig. 5.5.** Spestone Kaolin; remoulded at 100 per cent mc, uniaxial consolidation to 400 kPa; impregnated with Vestopal W; ultrathin section, vertical section, vertical cut, possibly inverted but otherwise correctly orientated; TEM. CV = direction of cut and vertical; H = horizontal. Most of the plate-like particles are seen edge-on; this is typical of vertical sections of consolidated samples. The particles are grouped in domains, in which all, or most, of the particles lie in approximately the same direction—local preferred orientation; the particles within the domains tend to be oriented horizontally—overall preferred horizontal orientation; and the domains themselves are elongated and tend to be oriented horizontally—overall preferred horizontal orientation of a different type. The domains tend to be curved and to be moulded together without interdomain voids, so we would classify this as turbostratic domain structure, which is typical of dispersed kaolin; cf. Figs. 5.7–5.9. The principal artefacts seen in this section are some small horizontal splits in the top third of the illustration; a hole in the centre, from which some particles have been pulled; and some shattering of vertical or nearly vertical particles. Also, between the arrows, there are two knife marks, which may just be evident as bands or lines of small holes parallel to the direction of cut, i.e. vertical. A full examination of the artefacts in such a section would require an original micrograph which had been printed on very 'soft' paper to give a very soft contrast (see Smart, P. (1966). Soil structure, mechanical properties and electronmicroscopy. Ph.D. thesis, University of Cambridge. Fig. 146). There has also been a little compression in the direction of cut, vertically; but the conclusions were verified by examining both vertical sections cut horizontally and horizontal sections. (From Smart 1966.) G.1071, Cambridge, BUR.C.

Fig. 5.6. Submarine sediment from Mississippi Birdfoot Prodelta from 120.4 m depth below sea floor under 70 m water; undisturbed sample; an almost vertical ultrathin section; otherwise preparation as Fig. 2.14. Note the strong but incomplete preferred orientation. The particles probably lie perpendicular to the stress direction. This structure was attributed to natural consolidation. (From Bennett *et al.* 1977.) R. H. Bennett of NOAA, Texas A & M Univ., McClelland Boring B-1.

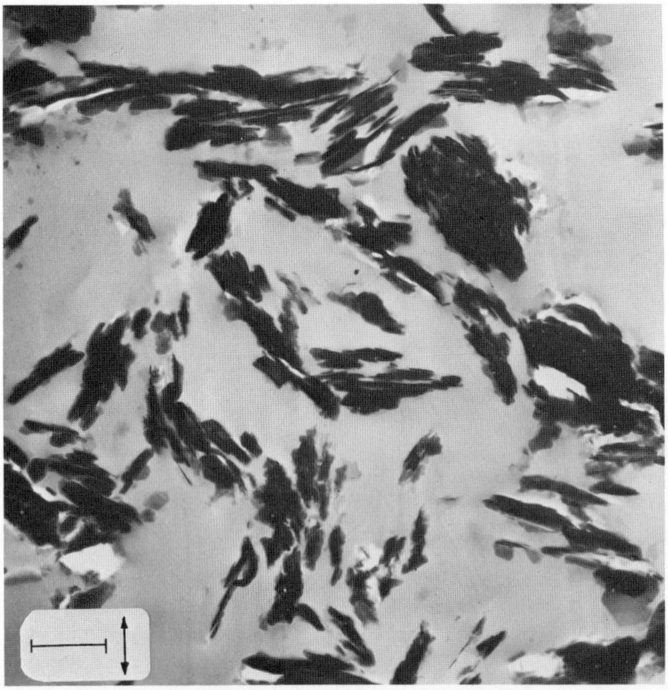

5.7

5.8

5.9

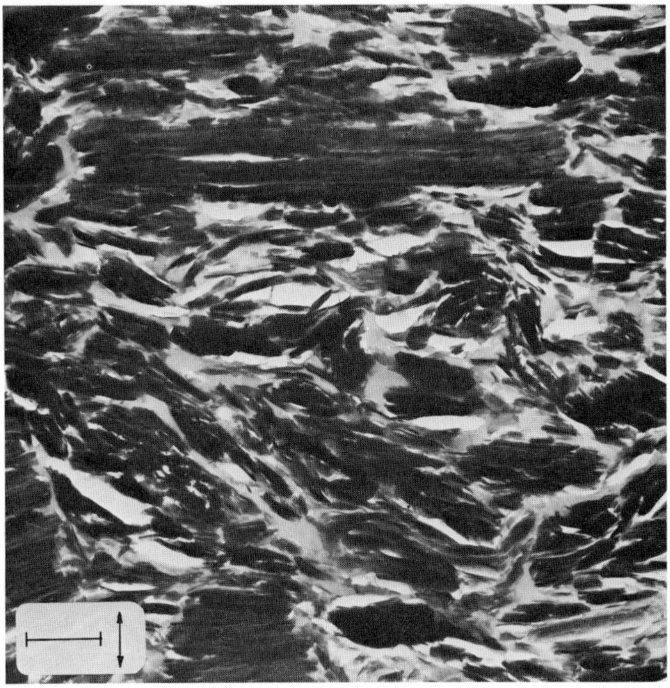

Figs. 5.7 5.9. SPS Kaolin; remoulded at 250 per cent mc, consolidated uniaxially to 0.001, 10, and 1000 kg cm^{-2} (0.1, 1000, and 100 000 kPa) respectively; allowed to swell; acetone, Vestopal W; ultrathin sections, vertical sections, vertically cut; TEMs. The arrows indicate the original vertical, except that the images may have been inverted top for bottom. **Fig. 5.8** illustrates discrete domain structure, in which the plate-like particles are arranged face-to-face in distinct domains, with distinct inter-domain voids; this is typical of flocculated kaolin, although a few single particles are seen at lower pressure in **Fig. 5.7.** In **Fig. 5.8,** the domains are outlined as a preliminary to measurement. During consolidation, at low pressures, the domains slipped over each other into sub-horizontal positions; at higher pressures, it was inferred from the measurements that the domains were themselves compressed. (From McConnachie, I. (1971). Electron microscopy of the consolidation of a kaolin. Ph.D. thesis, University of Glasgow, McConnachie 1974.) 1123, 469, and 955, I. McConnachie, Glasgow.

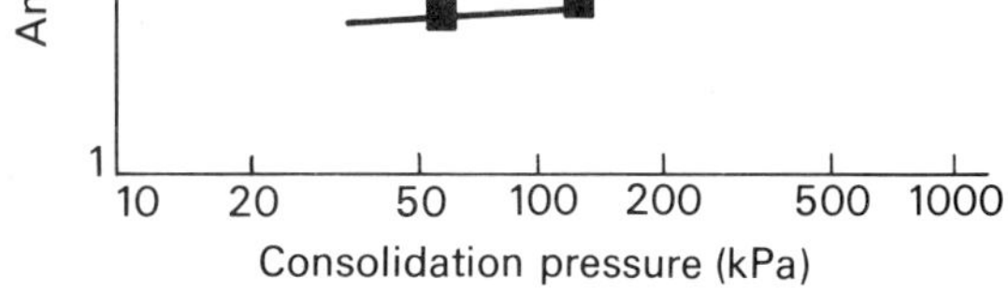

Fig. 5.10. Anisotropy of kaolin after consolidation to various pressures, determined from optical diffraction patterns from SEMs. Anisotropy ratio = 1 if isotropic. (From Tovey and Wong 1977.)

6 Domain structures

This Section illustrates consolidation in nature; however, in nature, particles do not merely slip, they also recrystallize. In some studies at least, the increase in consolidation owing to burial has been compounded with an increase in deformation (see Section 7) owing to tectonic activity; and in one or two of the examples, there may have been some heating in addition. Also, of course, recrystallization can occur without pressure. Fig. 6.1 is included here for comparison.

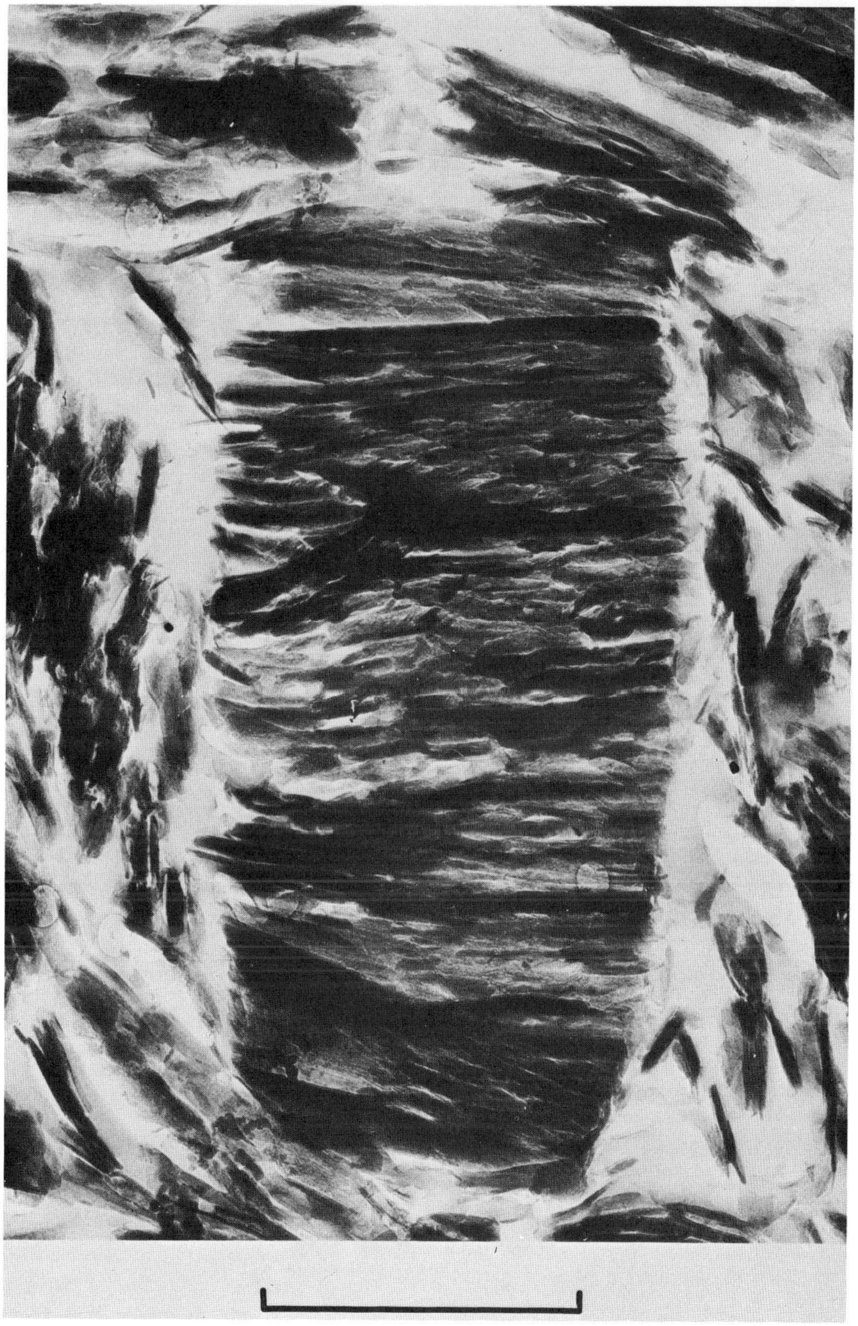

Fig. 6.1. Spestone Kaolin; remoulded at 100 per cent mc, uniaxial consolidation to 400 kPa, sheared rapidly; impregnated with Vestopal W; ultrathin section; vertical section perpendicular to the direction of shear, final orientation unknown; TEM. The micrograph shows a stack. Note the regular arrangement of the layers. This is a particle and neither packet nor domain. (From Smart, P. (1966). Soil structure, mechanical properties and electronmicroscopy. Ph.D. thesis, University of Cambridge.) G.970, Cambridge, AUS.S2.

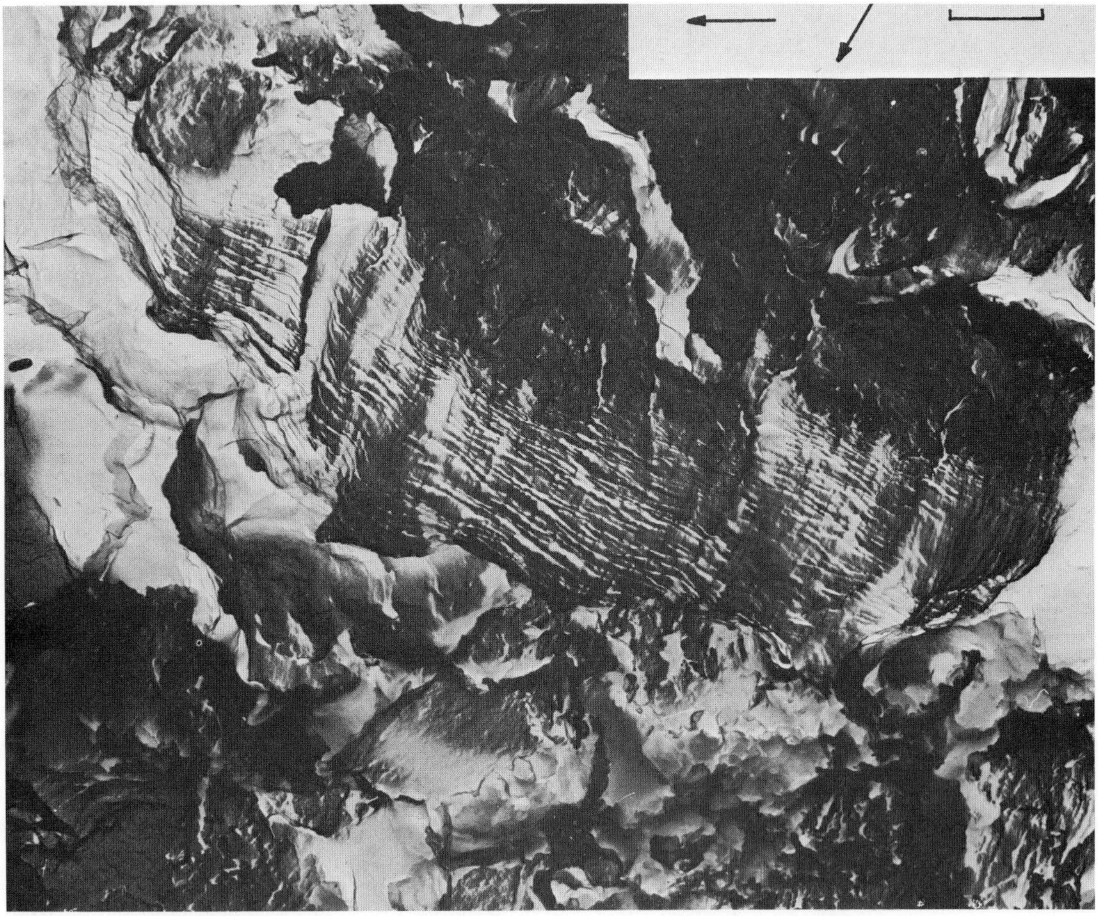

Fig. 6.2. This micrograph is the original illustration of domain or turbostratic structure of clay. Note that the crystal sheets are not only orientated and stacked but also interleaved across a considerable distance. Distortion of the sheets was attributed to stresses set up during drying. Technical details: somewhat degraded illite, K_2O content 4 per cent, from the B-horizon of a solodized solonetz in Willalooka, S. Australia; made homoionic; coarse particles removed; air dried; ground; equilibrated at relative vapour pressure = 0.96; compressed at 120 000 kPa (1200 atm.), rewet, and dried; replica obtained by a polyethylene technique using uranium shadowing from two directions, which are indicated by arrows; TEM. Supplied by L. A. G. Aylmore. (From Aylmore and Quirk 1960*a*; see also 1959, 1960*b*.) D. M. Hall.

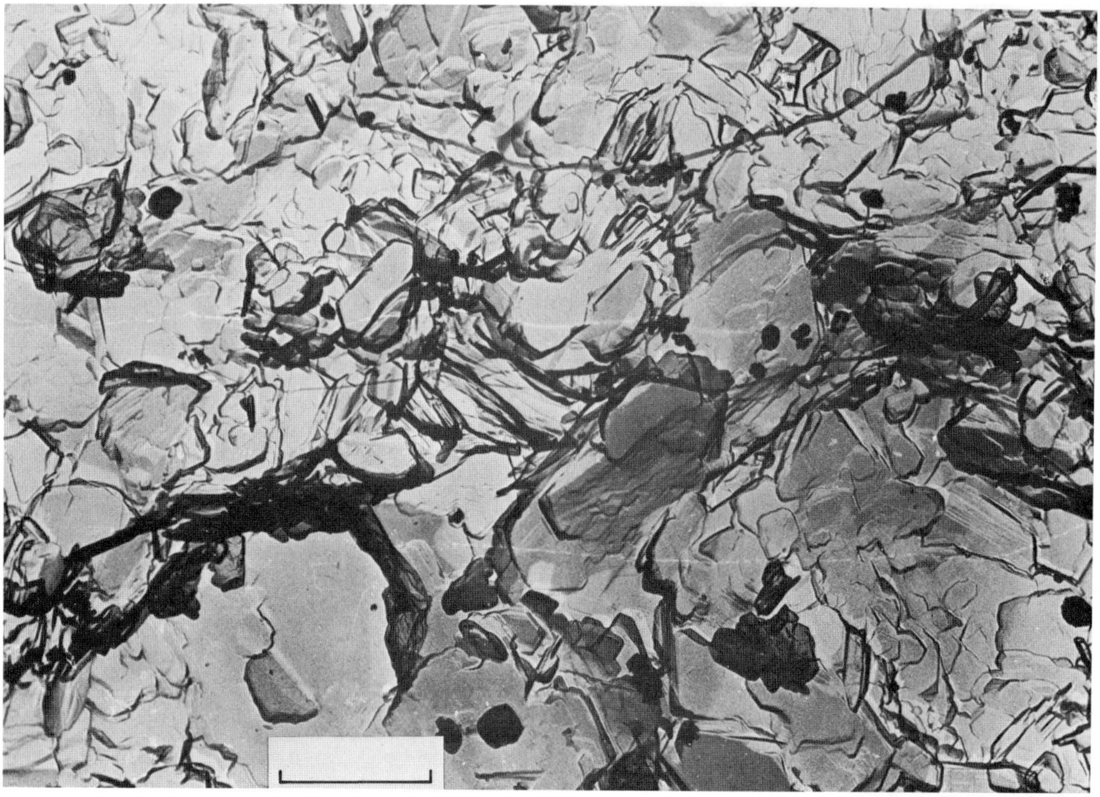

Fig. 6.3. Flint clay, kaolinite, from Beuker Pit, Owensville, Missouri; fracture surface; carbon–metal replica; TEM. Note how the crystals of kaolinite tend to merge into each other; we would classify this as an intergrown domain structure. Supplied by W. D. Keller. (From Keller 1968.) Charles B. Roth.

Figs. 6.4–6.7. Keller (1978) discusses five grades of transported kaolin: ball clay, which is exceedingly plastic; plastic fire clay; semi-plastic fire clay; semi-flint fire clay; and flint clay, which is hard and brittle. We show three of his SEMs illustrating the end and middle members, together with another included here for convenience.

Fig. 6.4. Ball clay from near Mayfield, Kentucky. Note the platy aggregates of tiny flakes face-to-face with a swirl pattern and the absence of large voids; we would classify this as turbostratic domain structure. This structure results from the movements during deposition and settlement; and it results in the high plasticity.

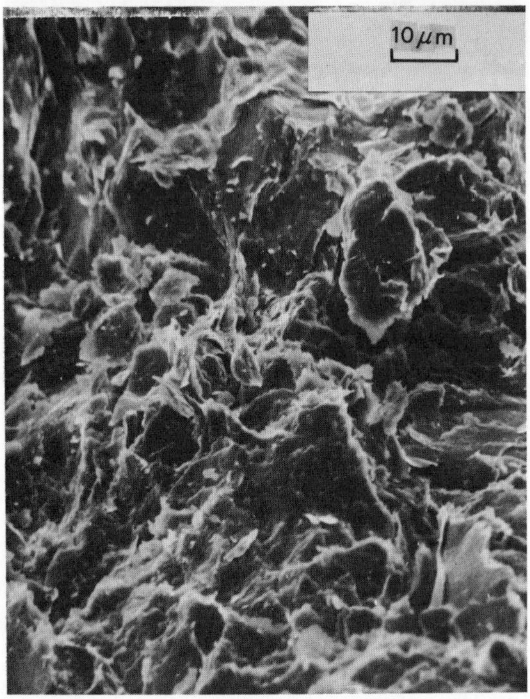

Fig. 6.5. Semi-plastic fire clay from Clearfield County, Pennsylvania. The structure and properties are intermediate between ball and flint clays.

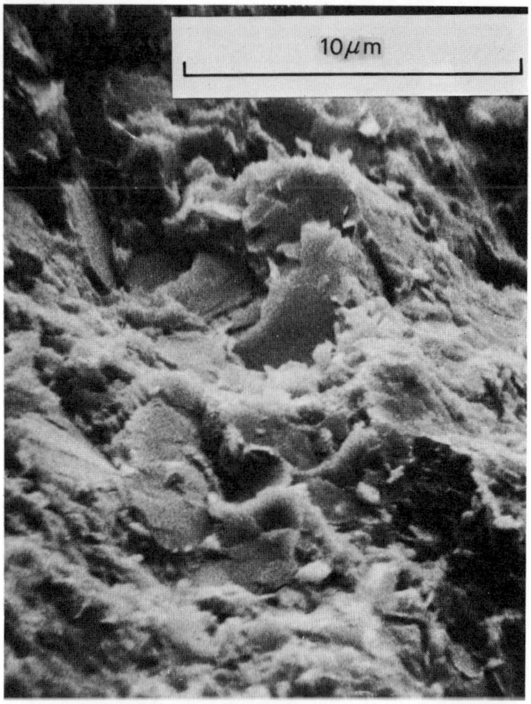

Fig. 6.6. Flint clay from Vereeniging, South Africa. The flakes have intergrown to form tightly interlocked units and a compact, dense, hard material.

Fig. 6.7. Residual, hydrothermally altered kaolin from Itaya Mine, Japan. Note the individual flakes or sheaves, but not stacks or books, probably best classified as close-packed random structure. Keller's general conclusion, based on several samples, was that hydrothermally altered kaolins tend to have a tightly packed, dense structure of small crystals in thin packets or sheaves. All supplied by W. D. Keller.

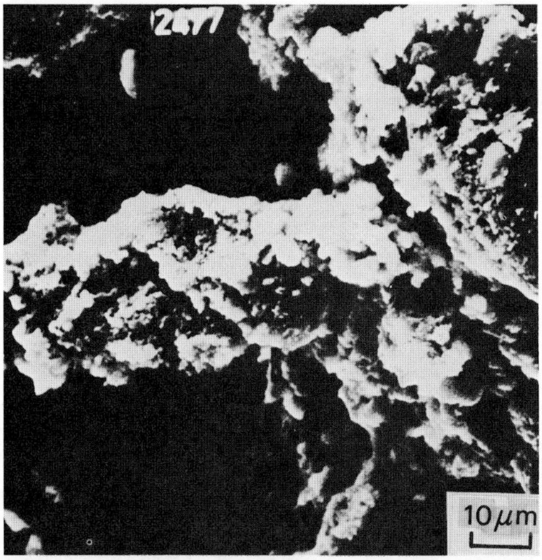

6.8

6.9

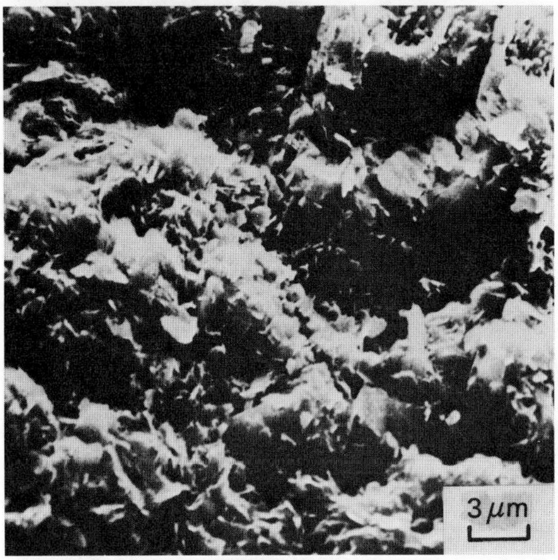

6.10

6.11

6.12

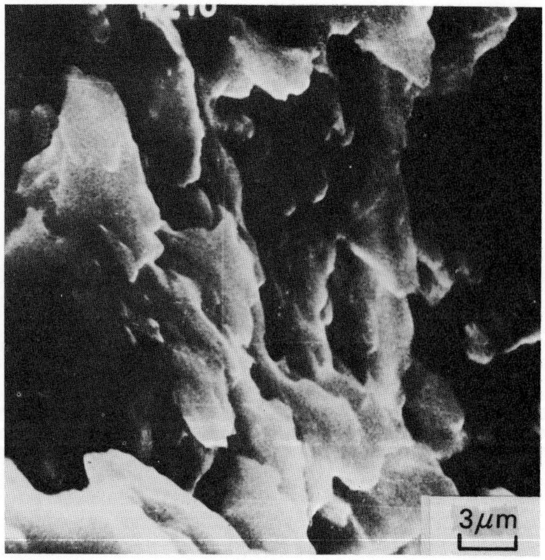

6.13

◁ **Figs. 6.8–6.13.** Karstic bauxites; undisturbed samples; fracture surfaces; C, Au–Pd; SEMs. Supplied by G. Bárdossy. (From Bárdossy, Csanády, and Csordás 1978.) Hungarian Aluminium Corp.

	Age	Principal minerals	Overburden (m)	Tectonic effect	Porosity (%)	Grain size (μm)	Aggregate size (μm)	Macro structure	Micro structure
6.8 Maré Island, Loyauté Islands	subrecent	a, gi	—	—	65	0.15	150	pelito-morph	'stacky' with macropores
6.9 South Manchester plateau, Jamaica	Upper Miocene	gi, h	—	—	45	0.2	15	pelito-morph	'stacky' with micropores
6.10 Iharkút mine, Hungary	Paleocene	k, g	10	weak	22	0.2	—	pelito-morph	uniformly micro-porous
6.11 Megara mine, Greece	Albian	b, h	50	strong	7	2.6	—	round-grain	uniformly compact
6.12 Chalkidiké peninsular, Greece	Kimmeridgian	c, d	50	strong and medium	2.5	10	—	micro-gran.	'crystalline-webby'
6.13 As 6.9, but subject to flooding; porosity is less; grain size is larger.									

Minerals: a amorphous; b boehmite; c chamosite; d diaspore; g goethite; gi gibbsite; h hematite; k kaolinite.

Figs. 6.14–6.16. Using both SEM and replicas, O'Brien (1970) related the degree of fissility of shales to the perfection of orientation of the clay plates comprising them. We show three of his TEMs of platinum palladium–carbon replicas of fracture surfaces; these are typical of a more extensive body of earlier American work which led to the concept of an inter-grown domain.

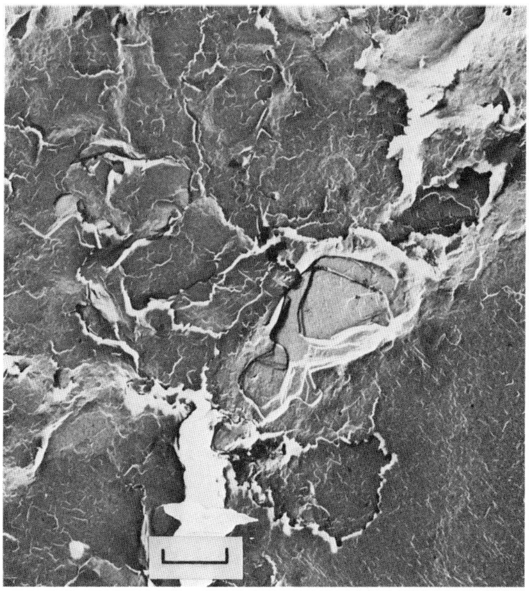

Fig. 6.14. Maquoketa Shale, Ordovician, from Illinois; abundant illite, some kaolinite; well developed fissility. The shadows are white; upstanding edges facing the source of shadowing metal, black; the prominent white feature near the bottom is probably a piece of the shale adhering to the replica. Note the high degree of orientation of the clay plates and the way in which they merge together. ▷

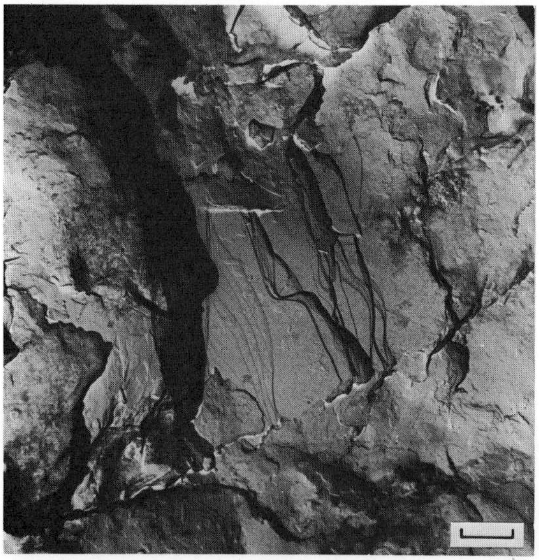

Fig. 6.15. Utica Shale, Ordovician, from New York State; abundant illite; well developed fissility. Recrystallization seems further advanced than in Fig. 6.14.

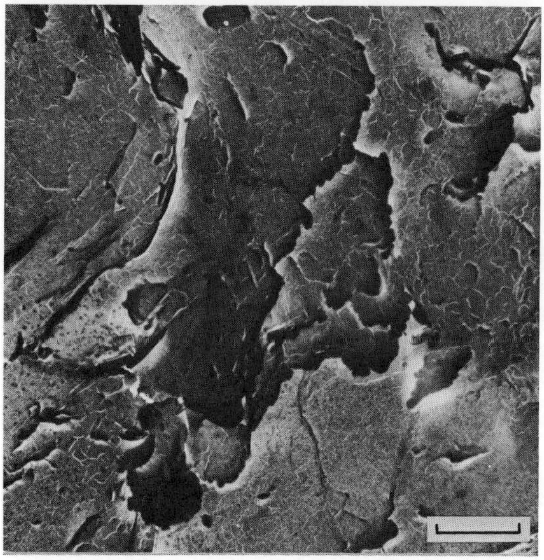

Fig. 6.16. Cuyahoga Formation shale, Mississippian, from Ohio; abundant illite and kaolinite, some mixed-layer clay; moderate to poor fissility. Note the merging of the clay plates and their local parallelism; although not too apparent from a monoscopic replica, many of the domains dip at various angles to the fissile plane.
All supplied by N. R. O'Brien.

Fig. 6.17. Purple slate, Middle Cambrian, from Nantlle, Caernarfon; ion-beam thinned, view parallel to lineation, trace of slaty cleavage approximately northwest–southeast; TEM. q, quartz; p, phyllosilicate, probably muscovite; h, probably hematite. Insert: selected area diffraction pattern of grain p; d-spacing ≈ 1 nm, characteristic of (002) of 2M muscovite. This structure, with its strong preferred orientation and almost complete space filling, is thought to have resulted from pressure solution facilitating mechanical rotation. Supplied by G. Oertel. (From Oertel and Phakey 1972.) P. P. Phakey, University of California, Los Angeles, W22. ▷

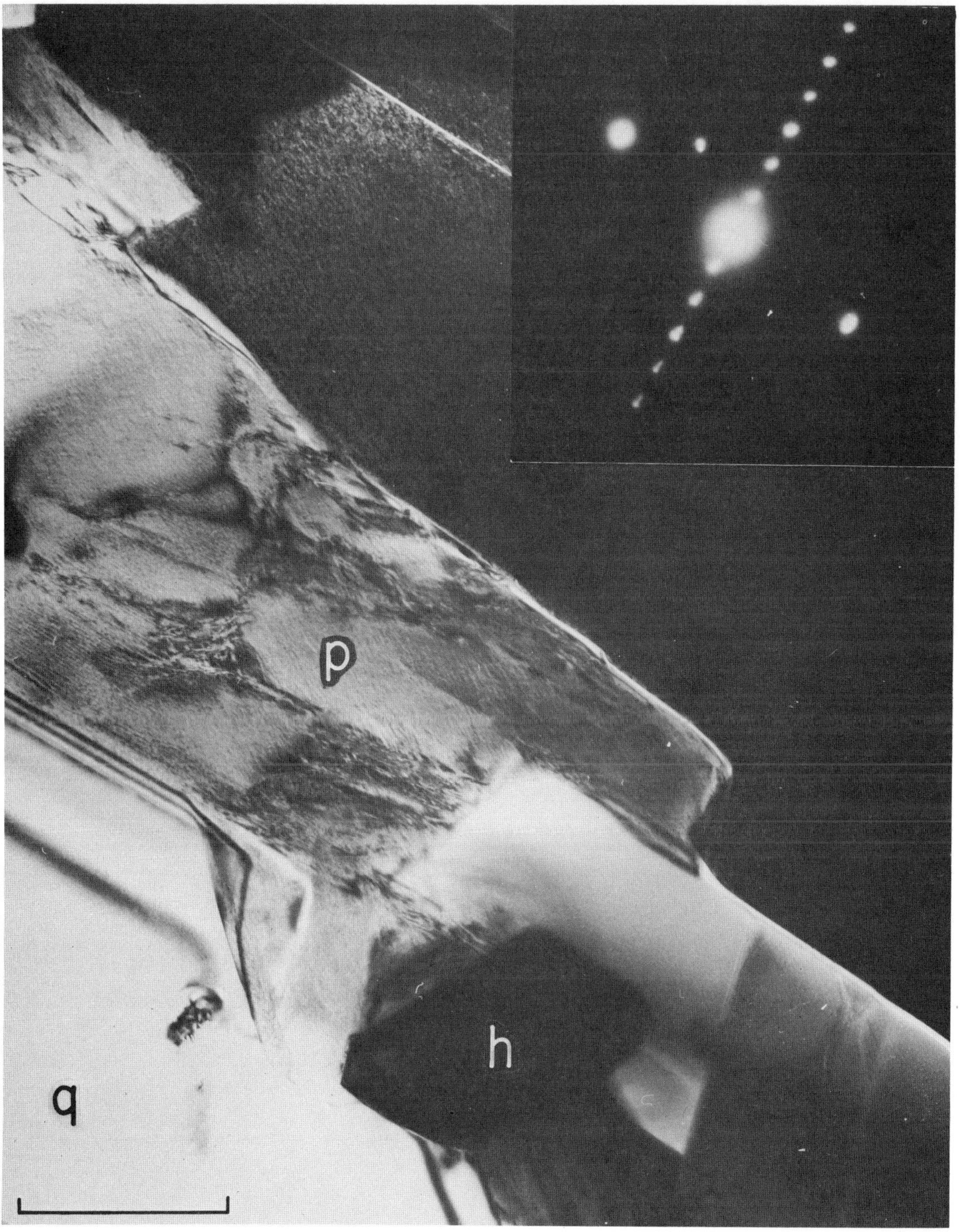
p
h
q

7 Deformation

The laboratory studies of consolidation illustrated above led naturally to similar studies of deformation, i.e. change of shape. These began later; and some important cases have not yet been examined. The principal experimental difficulty is that of collecting sufficient data to ensure that the work meets with the approval of statisticians. This is partly because it is difficult, even in laboratory studies, to obtain a uniform series of samples on which to experiment. For convenience, examples of both laboratory and natural deformation are included in this Section.

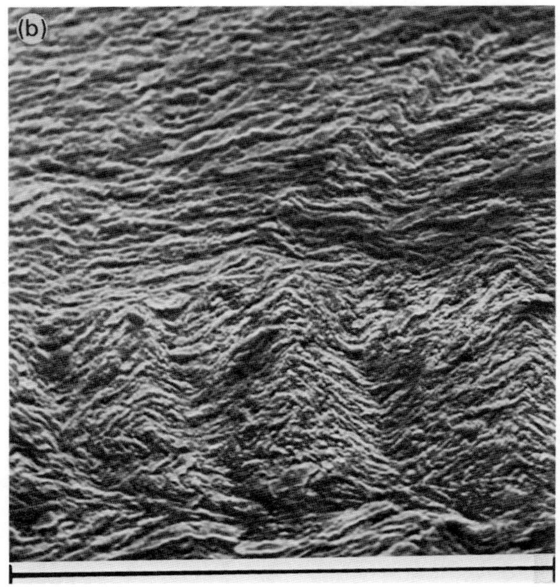

Fig. 7.1. Spestone Kaolin; remoulded, tested in a standard shear box; air-dried; vertical surface, peeled using '5-minute Epoxy'. (a) Montage of optical micrographs of part of the peeling showing bifurcation of the failure planes and microkinking between them. (b) Coated with 5 nm C, 20 nm Au; SEM, p 110 μm, θ 40°, showing the microkinking as herring-bone structure. Sample supplied by J. Cuckson. (From Wong and Tovey 1975.) B.1180.22, Cambridge.

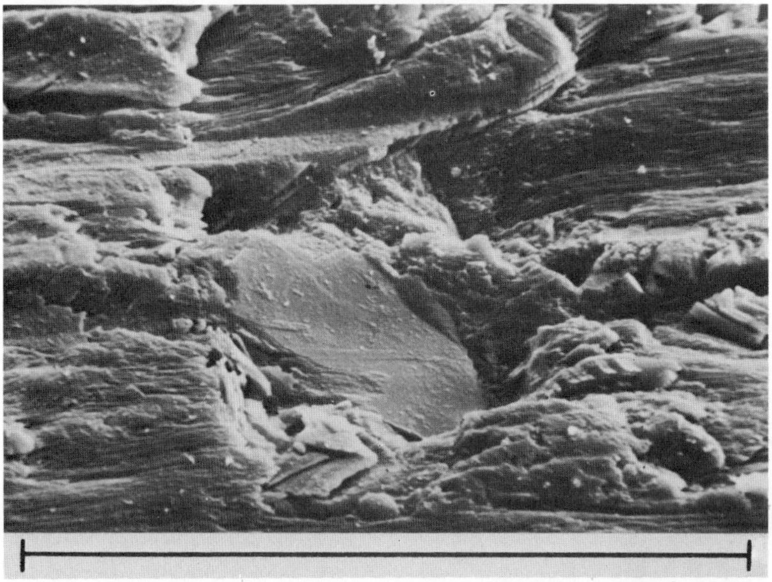

7.2

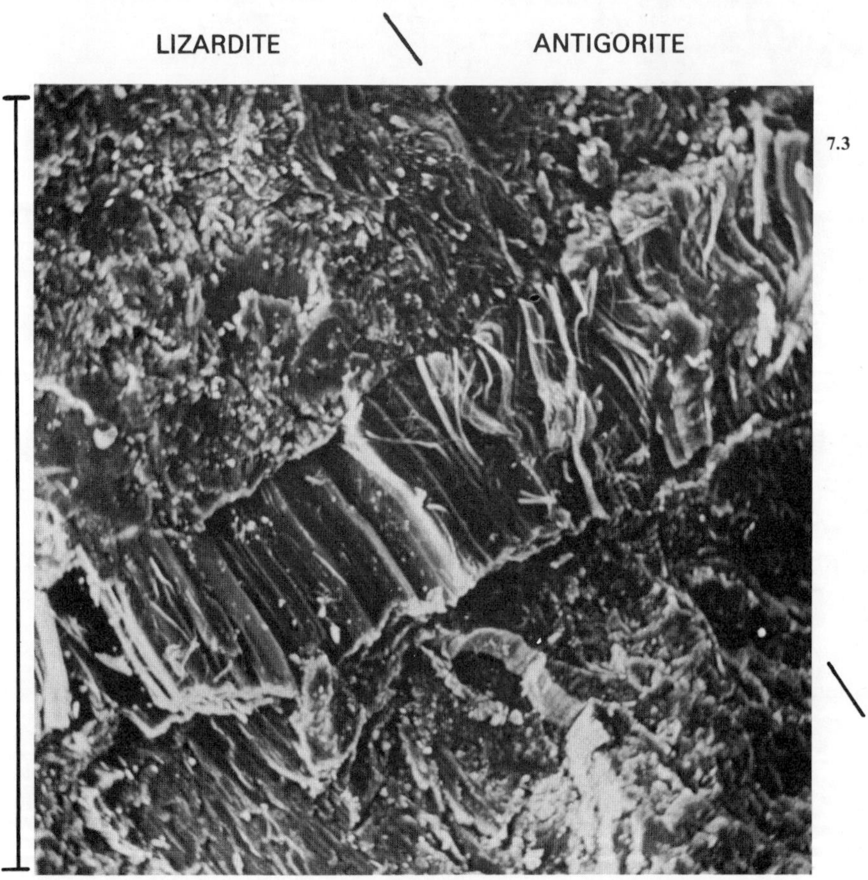

7.3

◁ **Fig. 7.2.** Chrysotile fibres forming on a lizardite plate, from Mount Maggiare, Cap Corse, Corsica; SEM, *p* 70 μm. The sample is in the early stages of shearing, some deformation being visible in the chrysotile. Taken and supplied by H. M. Prichard, Open University.

◁ **Fig. 7.3.** Chrysotile vein crossing antigorite and lizardite in dredged sample from Palmer Ridge; SEM, *p* 1 mm. The pattern of deformation of the vein suggests that the shearing is concentrated in the antigorite rather than the lizardite. (From Prichard 1979.) Taken and supplied by H. M. Prichard, East Anglia.

Fig. 7.4. Antigorite in dredged sample from Palmer Ridge; SEM, *p* 20 μm. This shows a well-developed shear zone in antigorite plates. (From Prichard 1979.) Taken and supplied by H. M. Prichard, Reading. ▷

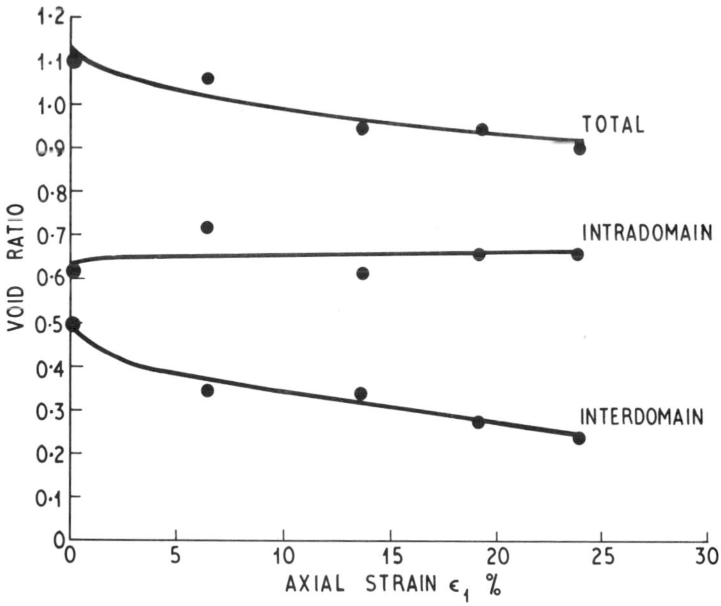

Fig. 7.5. Void ratios in normally consolidated SPS Kaolin during drained triaxial shear tests. The structure was rather similar to that shown in Fig. 5.8. Total void ratio from gravimetric analysis; inter-domain void ratio from ultrathin sections; intra-domain void ratio by difference. (From Smart and Dickson 1978.)

8 Failure

Work on the residual shear strength of soils led to a considerable interest in the structure of failure 'planes', which are perhaps more properly called failure zones especially in this context. In brief, if deformation is continued long enough, slickensided failure planes are formed and divide the soil into two or more parts, which may slide, one over the other, along the slickensided planes under very low stresses. Details of some failure zones are shown here from both laboratory and field studies. Samples tend to fall apart along failure planes, in which case it is impossible to get a complete cross-section of the failure zone. When this difficulty is avoided, it is often difficult to find the failure zone, especially when transmission electron microscopes are used, see De's work described in *Techniques*, Fig. 1.27 and Section 10.4.2 (Foster and De 1971).

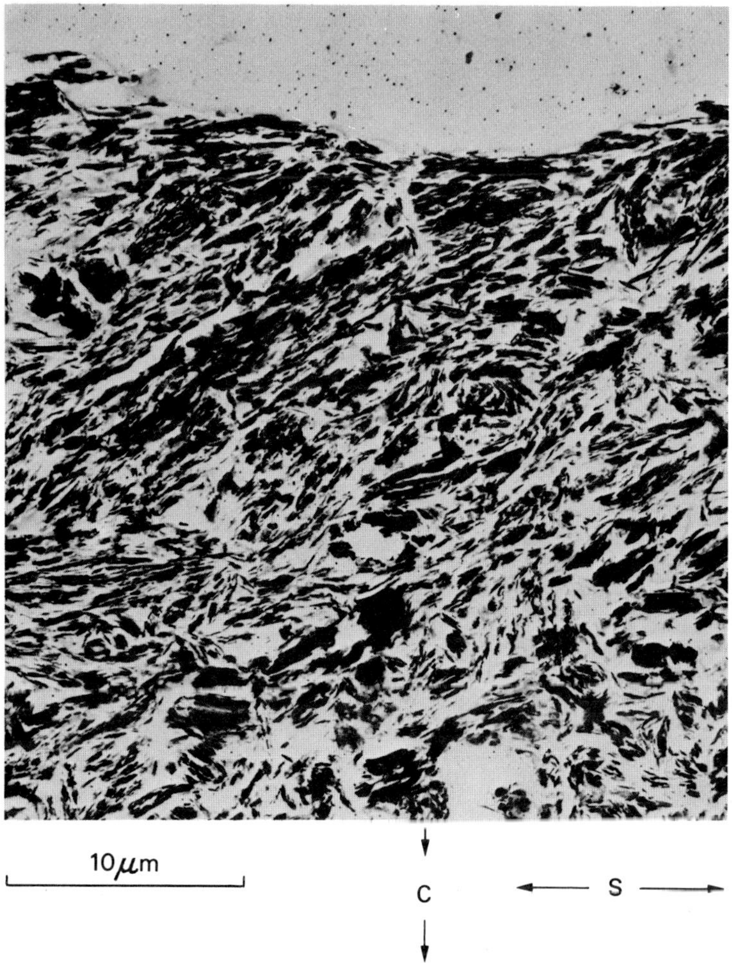

Fig. 8.1. Spestone Kaolin; remoulded at 136 per cent mc, consolidated to 560 kPa (80 p.s.i.), undrained triaxial shear test; air-dried; fractured along failure plane, which was slickensided; impregnated with Vestopal W; ultrathin section, section perpendicular to failure plane and parallel to direction of shear, cut perpendicular to failure plane as shown; TEM. C = direction of cut; S = direction of shear. Along the slickensided surface of the sample, there is a failure zone, in which the particles are aligned parallel to the zone; a thickness of approximately 1 μm of this zone is present on this half of the sample. Behind this surface zone, there is a considerable amount of preferred orientation at approximately 30° to the slickensided surface. The slickensided surface was orientated at approximately 66° to the horizontal when in the triaxial test machine. (From Smart, P. (1966). Soil structure, mechanical properties and electronmicroscopy. Ph.D. thesis, University of Cambridge.) G.4116, Cambridge, sample PS supplied by P. A. Loudon.

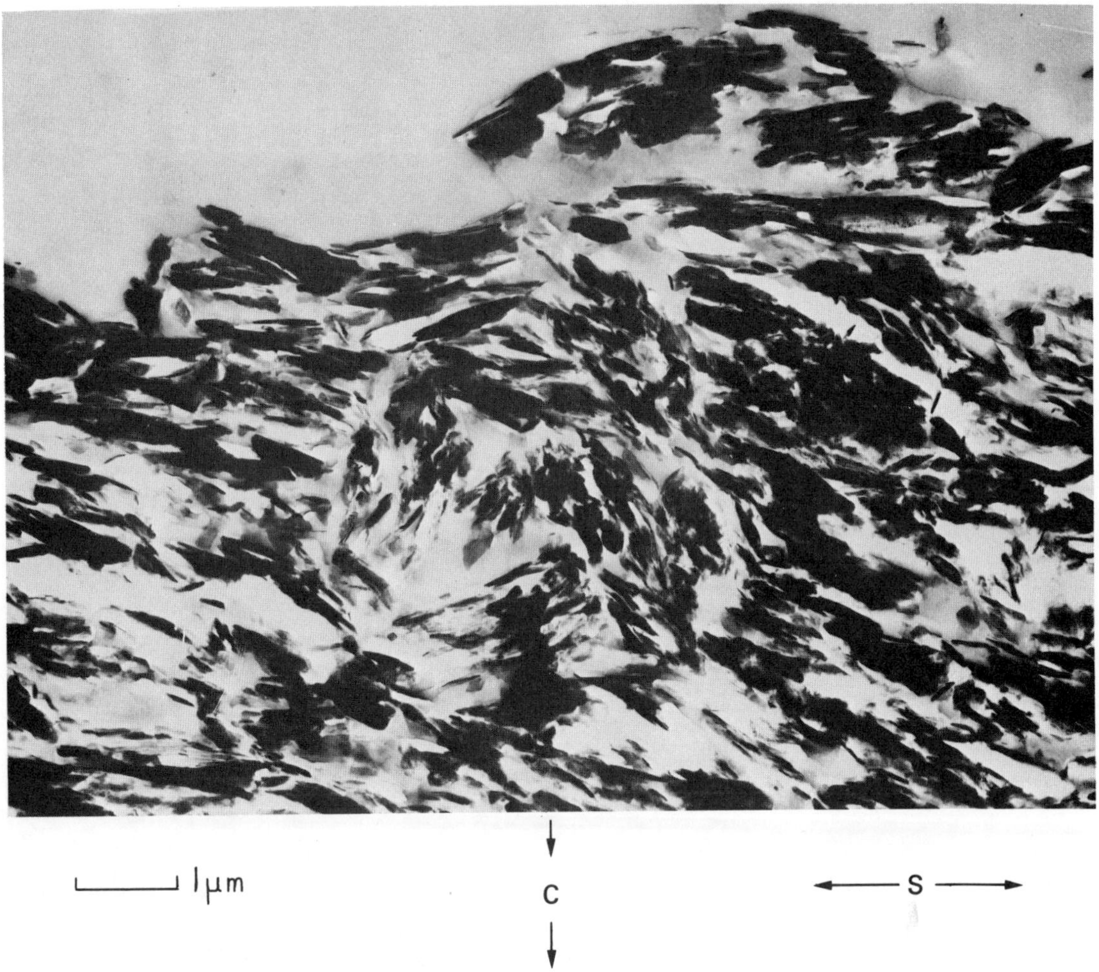

Fig. 8.2. From the same failure plane as Fig. 8.1; similar preparation. This micrograph shows a 'whorl', about 4 μm ϕ, in which the particles are arranged tangentially. The whorl is situated within a well-developed failure zone. (From Smart, P. (1966). Soil structure, mechanical properties and electronmicroscopy. Ph.D. thesis, University of Cambridge.) G.3471, Cambridge, sample PS.

Fig. 8.3. Spestone Kaolin; remoulded at 80 per cent mc, uniaxial consolidation to 400 kPa, sheared rapidly in a shear box; freeze-dried; self-parted along the failure plane; chromium shadowed at 45° elevation, two coats of carbon; replica; TEM; negative print. Note the striations, which resemble wear marks on alumina surfaces (King 1960), but which also resemble fibrous clay-mineral particles: fibrous minerals have not been mentioned in the analyses of this kaolin. (From Smart, P. (1966). Soil structure, mechanical properties and electronmicroscopy. Ph.D. thesis, University of Cambridge.) C. 30622, Cambridge, AUR.S.

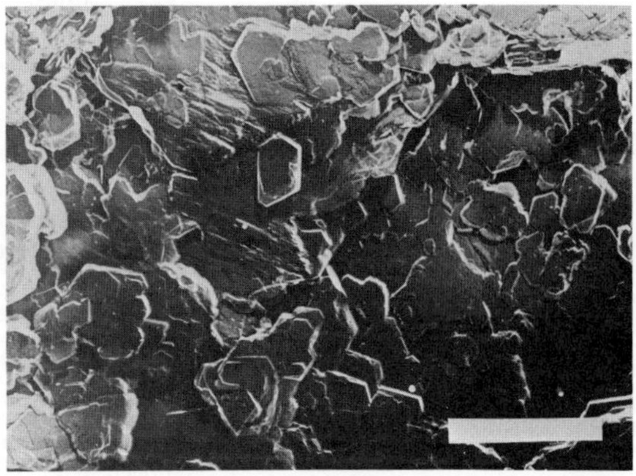

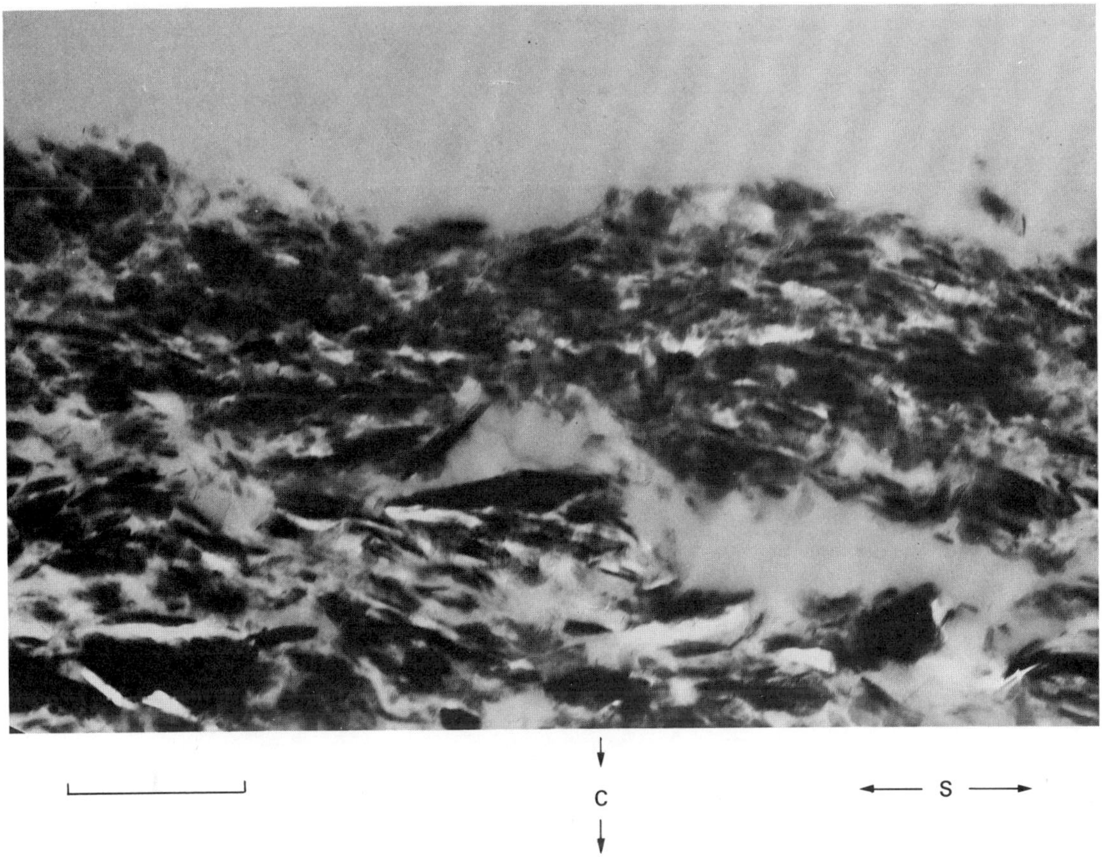

Fig. 8.4. From the same failure plane as Figs. 8.1 and 8.2, similar preparation. This micrograph is at a higher magnification than the preceding two. The particles in the failure zone are smaller than those seen in the other micrographs from this sample; the longest visible dimension of the smallest particle is approximately 20 nm. Also these particles are of irregular shape and indistinct outline. It is thought that these small particles were formed by comminution within the failure zone as a result of shearing during the triaxial test. The possibility that the particles were reduced in size when the section was cut should also be considered. Particles parallel to the direction of cut are often shattered during this cutting, however, the direction of cut was from top to bottom of the micrograph, and these particles tend to be perpendicular and not parallel to this direction. Also, although a little cutting damage is evident, the impregnating plastic is relatively intact. It is therefore thought that there was no serious damage during cutting in this case. Another possibility is that these particles are rogue particles, which were included in the sample *ab initio* by mistake. Elimination of this possibility would require more replication than was possible in the original study (From Smart, P. (1966). Soil structure, mechanical properties and electronmicroscopy. Ph.D. thesis, University of Cambridge.) G.3316, Cambridge, sample PS.

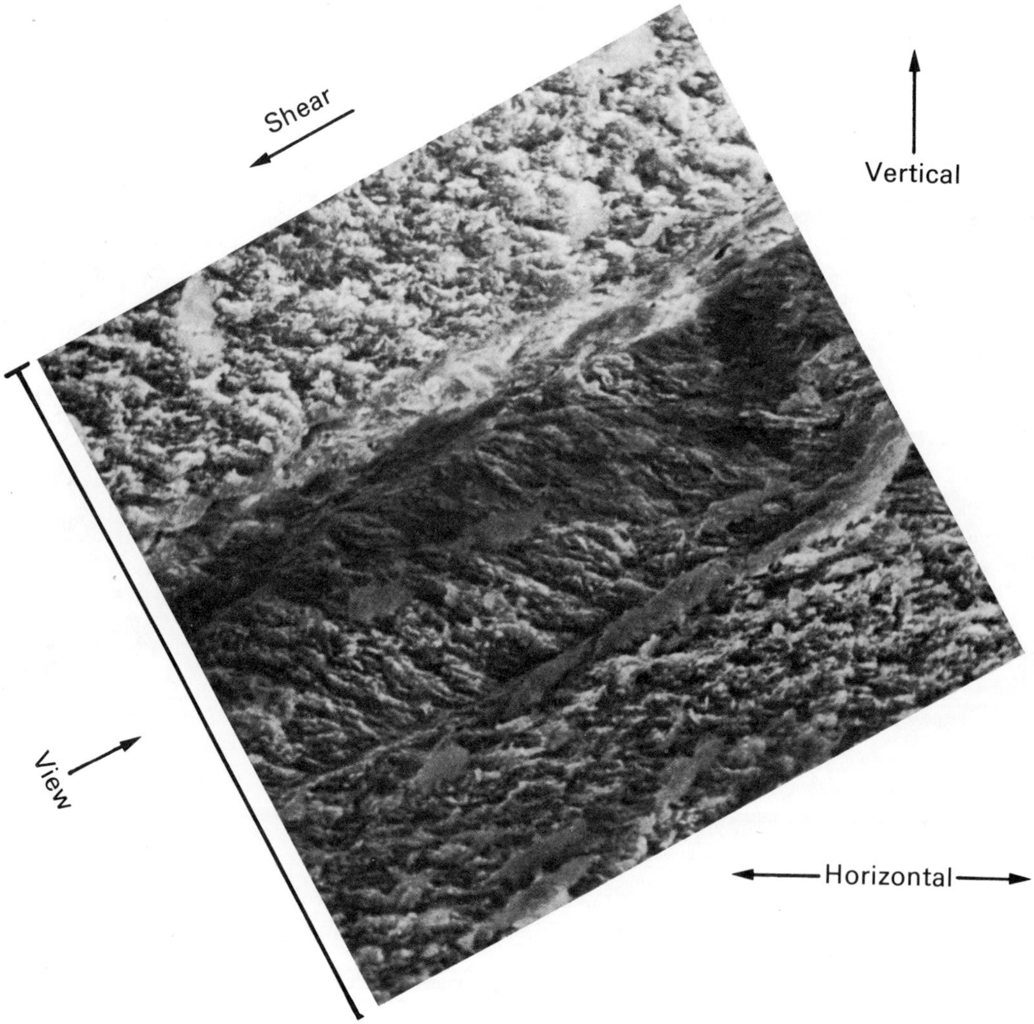

Fig. 8.5. Spestone Kaolin; remoulded at 160 per cent mc, uniaxial consolidation, model footing test; air-dried; vertical fracture; 5 nm C, 20 nm Au; SEM, *p* 208 μm, *θ* 40°. This was taken from beneath the footing. Two highly oriented failure planes divide the micrograph into thirds. In the lower third, the original bedding is seen. In the upper third, this has rotated a few degrees clockwise and is less clear. In the centre, preferred orientation has been developed by shearing; after allowing for tilt, this is at about 44° to the failure planes (From Wong, K. Y. (1975). Microfabric changes during the deformation of clays. Ph.D. thesis, University of Cambridge) B47.31A, Cambridge.

Fig. 8.6. Spestone Kaolin; remoulded at 160 per cent mc, uniaxial consolidation to 187 kPa, centrifuge test on model slope; air-dried; 10 nm C, 20 nm Au; montage of SEMs, *θ* 40°. Note the highly oriented failure zone. (From Tovey, N. K. (1971). A selection of scanning electron micrographs of clays. Cambridge University Engineering Department Technical Report CUED/C—SOILS/TR5.) 261.16 etc., Cambridge, LJE.A2.1A.1FZ.

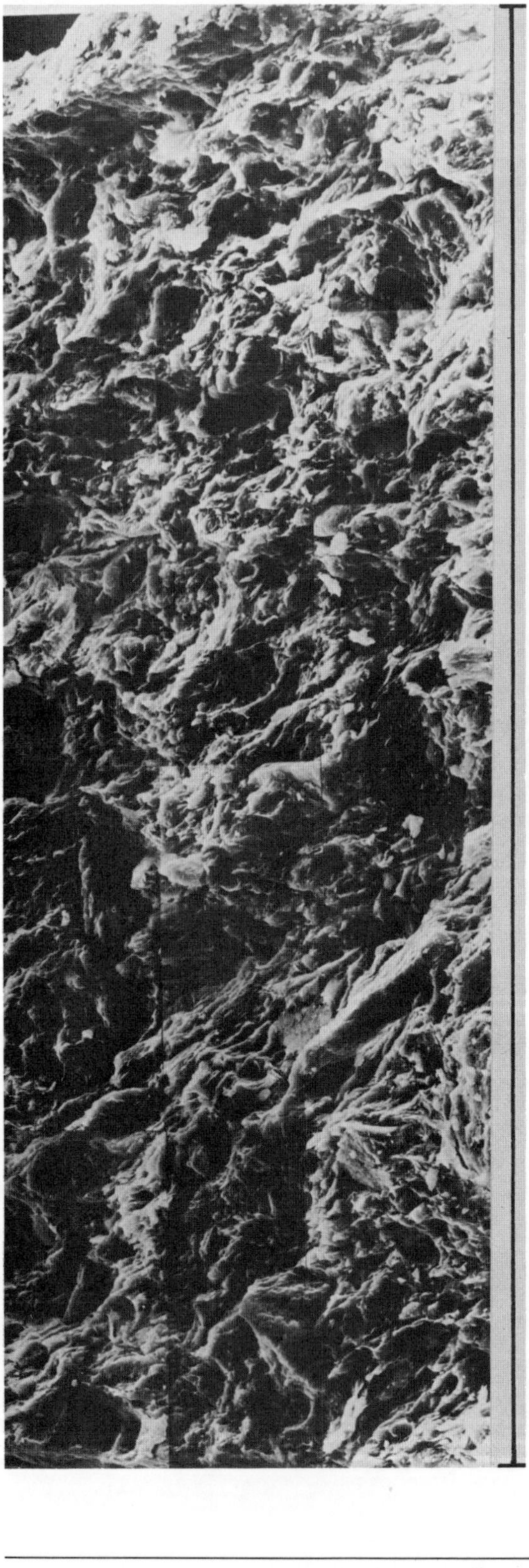

Fig. 8.7. Clay skin from a pile, which had been vibrated continuously during placement and then loaded to failure, London Clay; air-dried; montage of SEMs, vertical view of vertical fracture, i.e. vertical is up–down and horizontal is across, *p* 360 μm, *θ* 40°. The complete thickness of the skin is shown. The pile was on the right. The soil adhered to the pile; and failure occurred through the soil on the left, which now shows somewhat greater orientation. Sample supplied by M. J. Tomlinson. (From Tovey 1973.) Cambridge, ULON.MJT.DP6.

Fig. 8.8. As Fig. 8.7, except that this pile had been jacked into position; vertical view of vertical fracture, *p* 200 μm, *θ* 40°. This skin was 8 mm thick. Failure again occurred through the soil. This shows the failure zone. Note the higher degree of orientation; the ultimate load on this pile was 25 per cent less than that on the vibrated pile. ULON.MJT.JP5.

9 Organic matter

The theme of this Section is the interaction of the organic and inorganic elements in an ordinary, i.e. mainly mineralic, soil, or sediment. The organic elements include inorganic molecules, ions, enzymes, etc., which have usually been either not seen or not recognized in undisturbed samples; bacteria; other micro-organisms; larger fauna; roots, etc.; and debris. Almost all the work so far has concentrated on bacteria. These represent only a fraction of the organic matter, which itself constitutes only a fraction of an ordinary soil; so they are difficult to find. Most, but not all, of the work so far has been concentrated on the rhizosphere, a sort of no-man's land between root and soil. The other principal difficulty is that of ensuring and verifying that neither are the original bacteria destroyed nor are additional bacteria bred within the samples. Before attempting such work, remember too that some bacteria are lethal to electron microscopists.

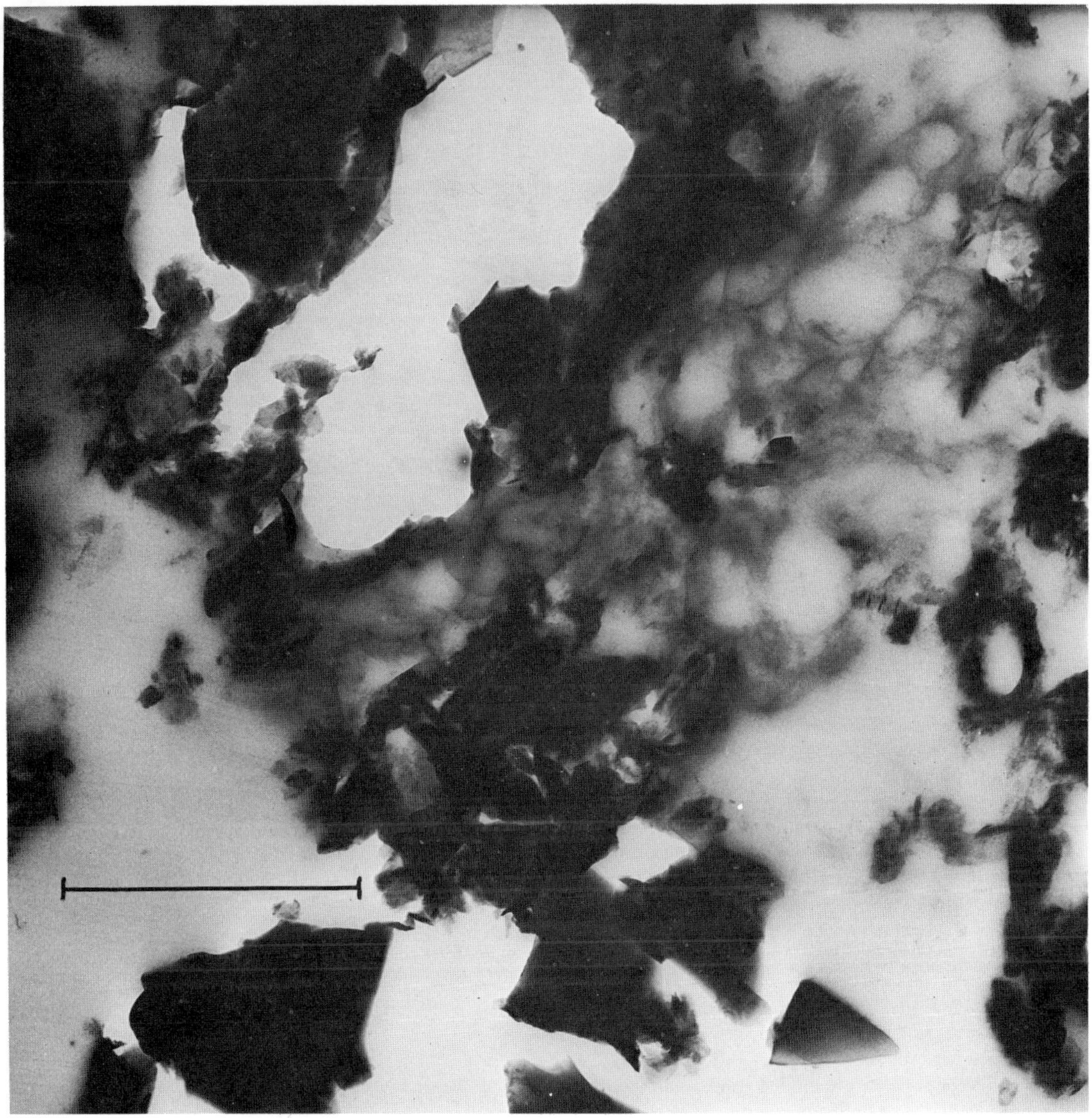

Fig. 9.1. Brackish-water, iron sulphide rich, organic clay from Morjärv, 6 m depth; undisturbed sample; impregnated; ultrathin section; TEM. Note the fibrous organic network in the upper right quarter of the micrograph, and the loop 0.2 μm ϕ of presumed bacterial origin half-way down the right-hand side. There are some holes, artefacts, in this section. Micrograph supplied by R. Pusch.

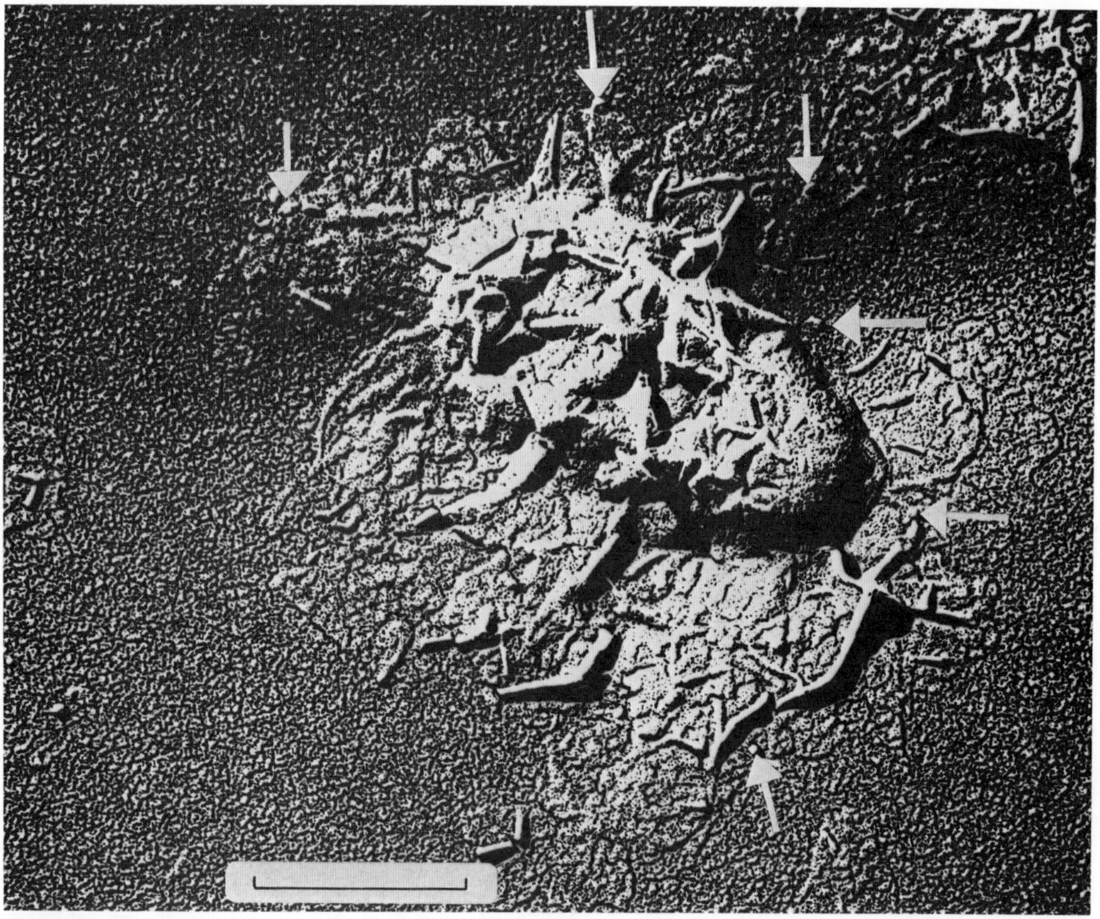

Fig. 9.2. *Escherichia coli* bacterium, montmorillonite, and bacteriophage virus, from suspension; freeze-dried; shadowed, 60 per cent Au 40 per cent Pd; TEM; negative print. The phage particles are minute spheres, arrowed. The montmorillonite coats the bacterium, thereby protecting it from attack by the bacteriophage. Supplied by K. C. Marshall. (From Roper and Marshall 1974.)

Fig. 9.3. Fen Clay from The Wash, i.e. virtually undrained ground, between the Old and New Bedford Rivers at TL 473855, 2.9 m depth; undisturbed sample, NGI sampler; alcohol, acetone, Vestopal W; ultrathin section, vertical section; TEM. This shows an occurrence of random or single-plate card-house structure, which is thought to have been formed by micro-organisms, which rearranged the plates to form loop-like structures 0.5–1 μm ϕ, cf. the following figures, thereby almost filling a pre-existing void which extends across the width of the micrograph. The liquidity index of this sample was about 1.5; but the general structure of the sample was less open than that shown here. (From Stirling, J. M. (1973). Structure of Fen Buttery Clay. B.Sc. report, University of Glasgow; Smart 1975.) 3468, P.S., Glasgow, 338.5.1.V.

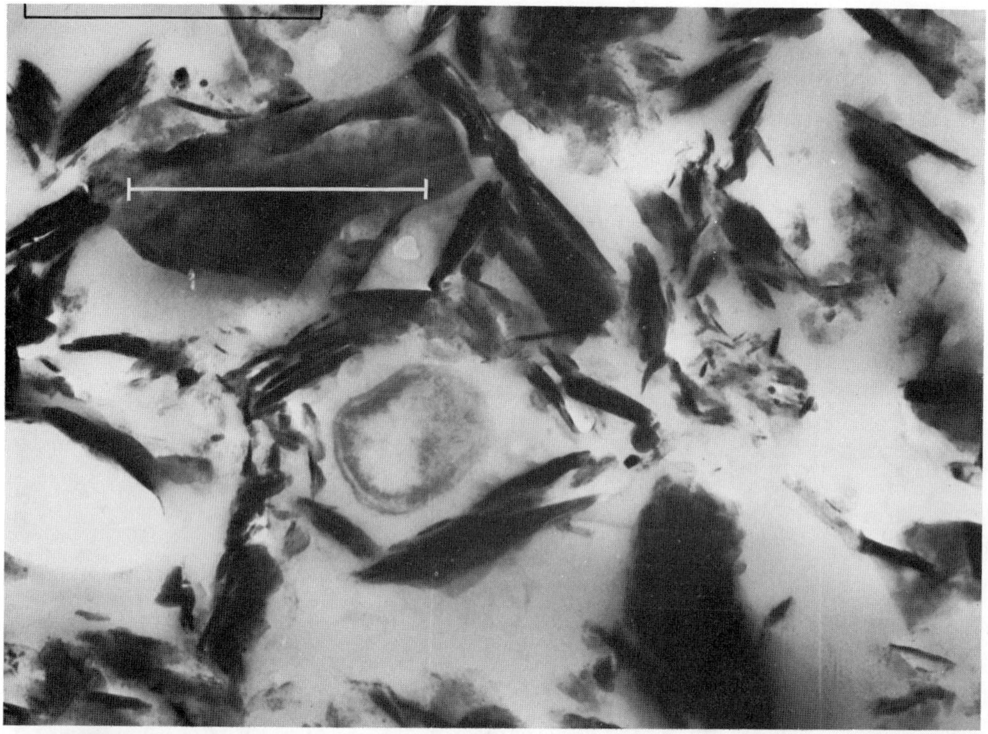

9.4

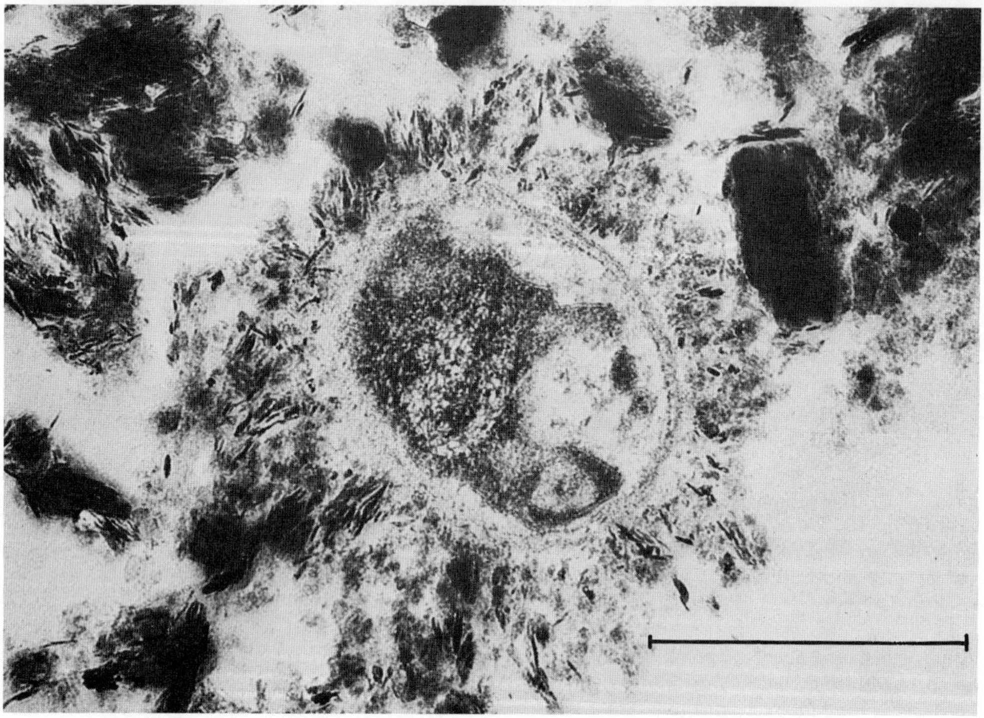

9.5

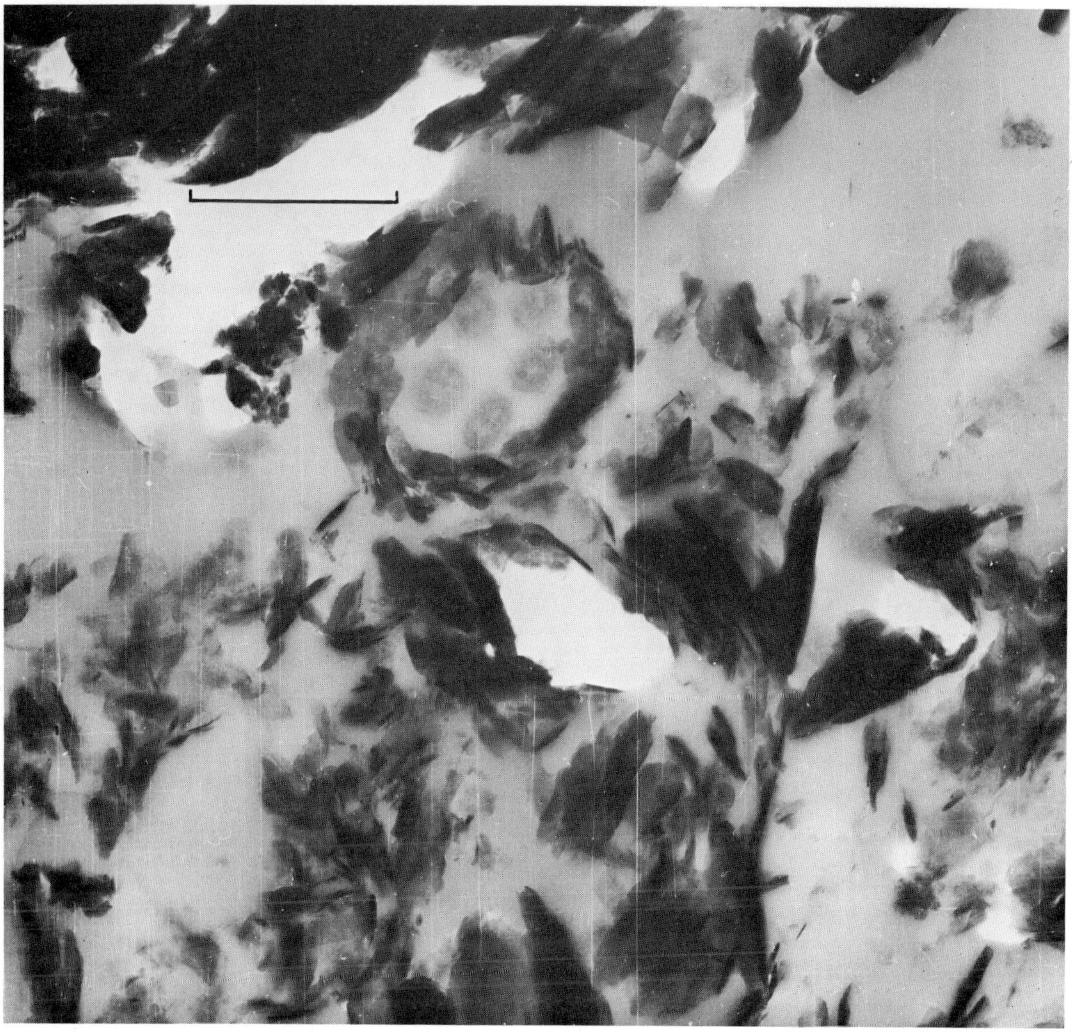

Fig. 9.4. Submarine sediment from the Severn Estuary; undisturbed sample from 40–50 mm below the top of the sediment under 9 m water; glutaraldehyde–gelatin, osmium tetroxide, uranyl acetate, Araldite CY212; ultrathin section, vertical section; TEM. This shows a cross-section of a bacterium or blue–green alga, 0.5 μm ϕ encased in clay plates, inner diameter of casing = 0.7 μm. It is thought that the space between the micro-organism and the casing may be filled with polysaccharide, although this has not been confirmed. The original micrograph seems to show some very faint threads extending from the micro-organism into this space. This micro-organism was thought to be dead and beginning to decompose at the time of fixation, although it was difficult to be certain. Sample supplied by the Institute of Oceanographic Sciences, Taunton. 6909, P. A. Maclean, Glasgow, Core No. 293.

Fig. 9.5. Mallee Loam from a wheatfield in S. Australia; undisturbed sample taken in a metal cylinder and subdivided; fixed: agar, acrolein–glutaraldehyde in a phosphate buffer, washed, osmium tetroxide; embedded; ultrathin section; TEM. This shows a bacterium inside a casing of very fine clay plates, which in this case tend to be arranged radially. Micrograph supplied by R. C. Foster.

△

Fig. 9.6. Details as Fig. 9.4, except that the depth was 30–40 mm below the top of the sediment, and only osmium tetroxide was used for staining. This shows a group of bacteria, each measuring approximately 0.2 μm ϕ within a casing of clay 0.7 μm ϕ. The vertical white lines are photographic defects. 7597, P. A. Maclean, Glasgow, Core No. 293.

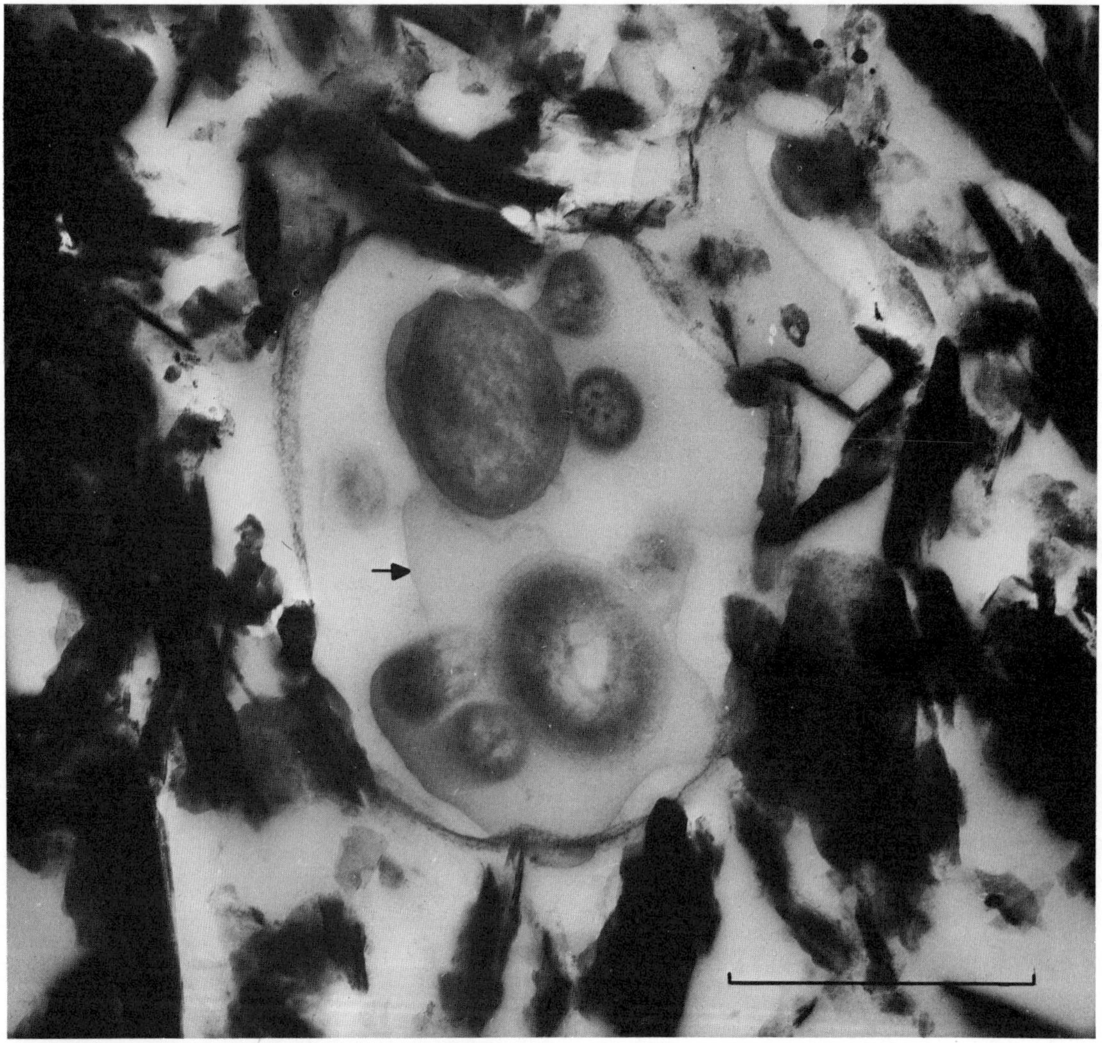

Fig. 9.7. Details as Fig. 9.4, except that the depth was 30–40 mm below the top of the sediment. This shows a group of bacteria or blue–green bacteria, gram-negative, 0.5–0.2 μm ϕ, which are thought to be cross-feeding. The contrast change, arrowed, is thought to be a water mark, i.e. an artefact. 6947, P. A. Maclean, Glasgow, Core No. 293.

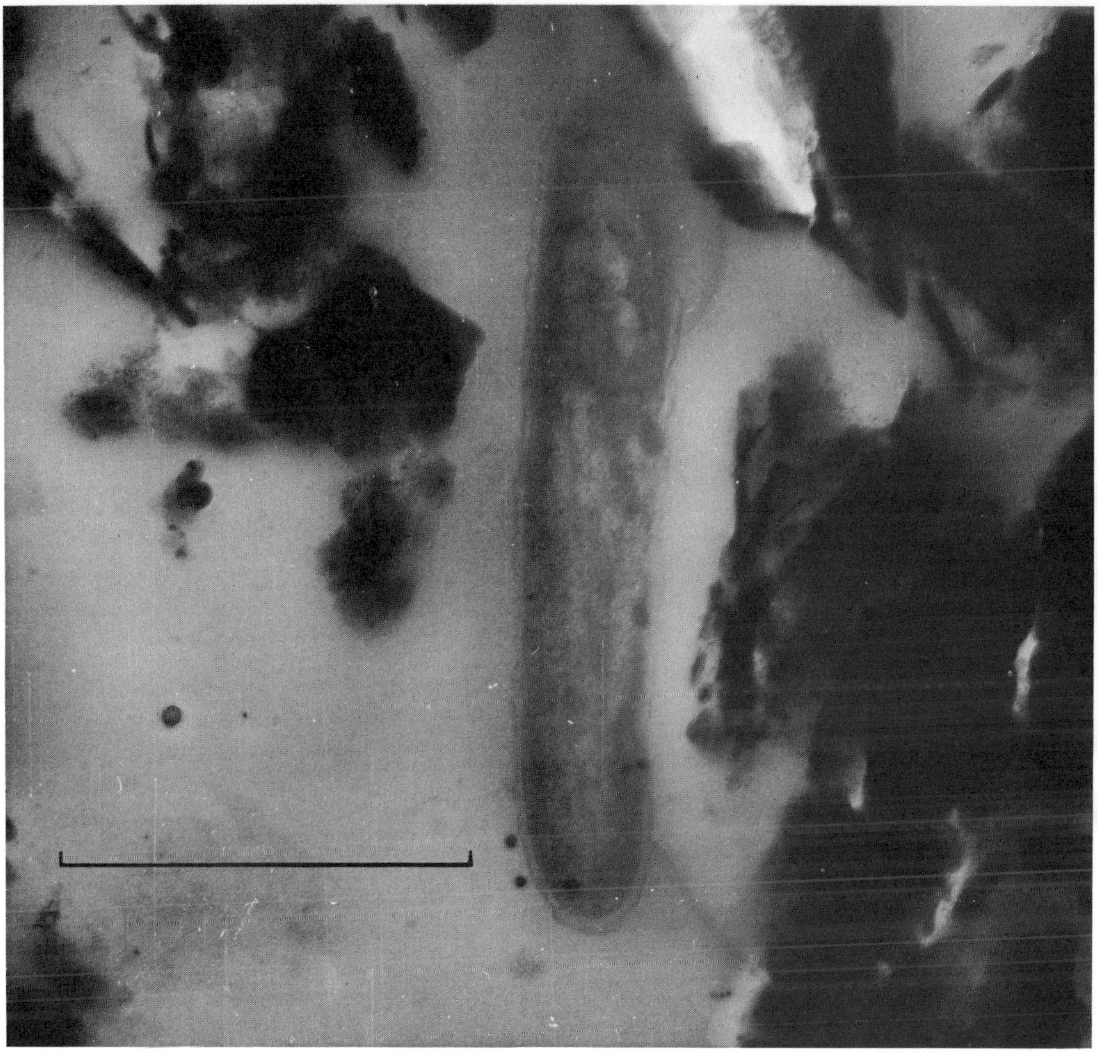

Fig. 9.8. Details as Fig. 9.4, except that only osmium tetroxide was used for staining. This ultrathin section of an undisturbed sample of sediment shows a bacterium or blue–green alga, probably gram-negative, 2 μm long × 0.3 μm ϕ, which is thought to be swimming freely in a void in the sediment. The small dark circular marks near the scale mark are photographic defects. 6968, P. A. Maclean, Glasgow, Core No. 293.

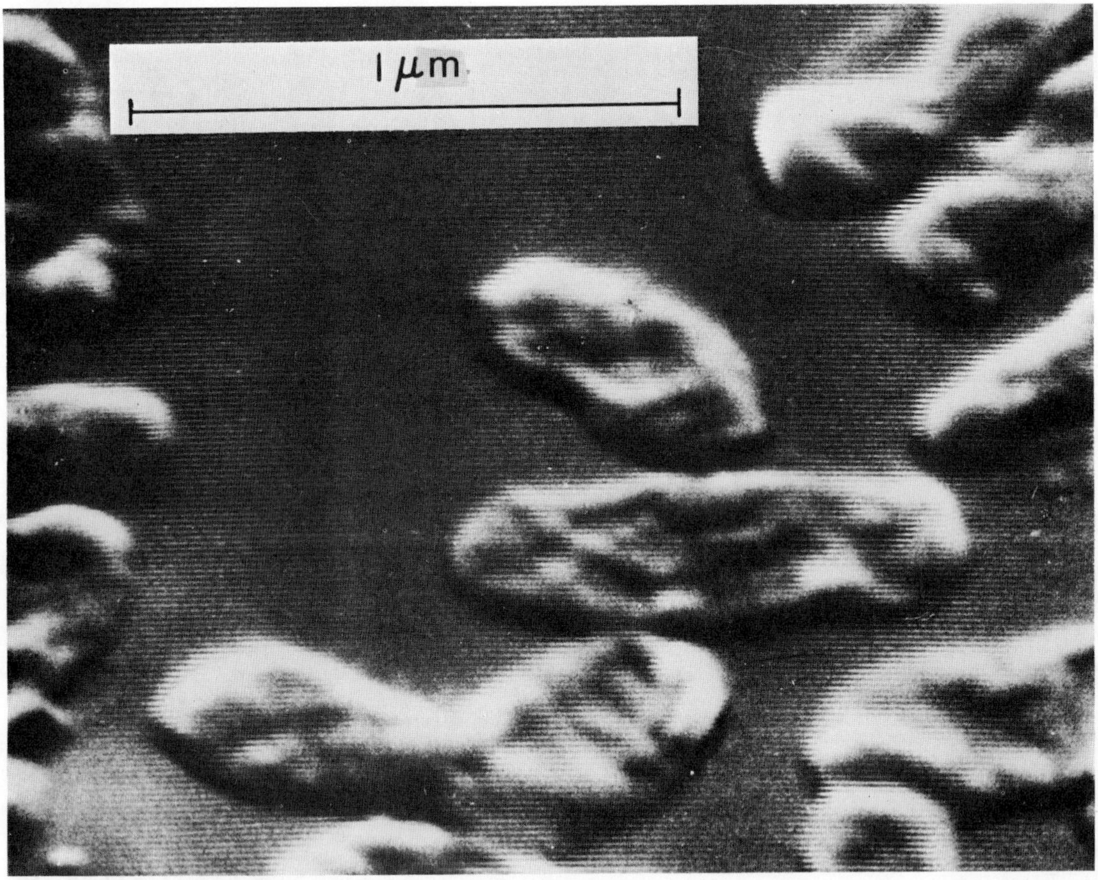

Fig. 9.9. SEM of *Ferrobacillus–Thiobacillus* bacteria isolated from chemically altered and heaved Billings Shale from beneath the Bell Canada Building, Ottawa. The bacteria are believed to have contributed to the alteration. Supplied by E. Penner. (From Penner, Gillott, and Eden 1970.)

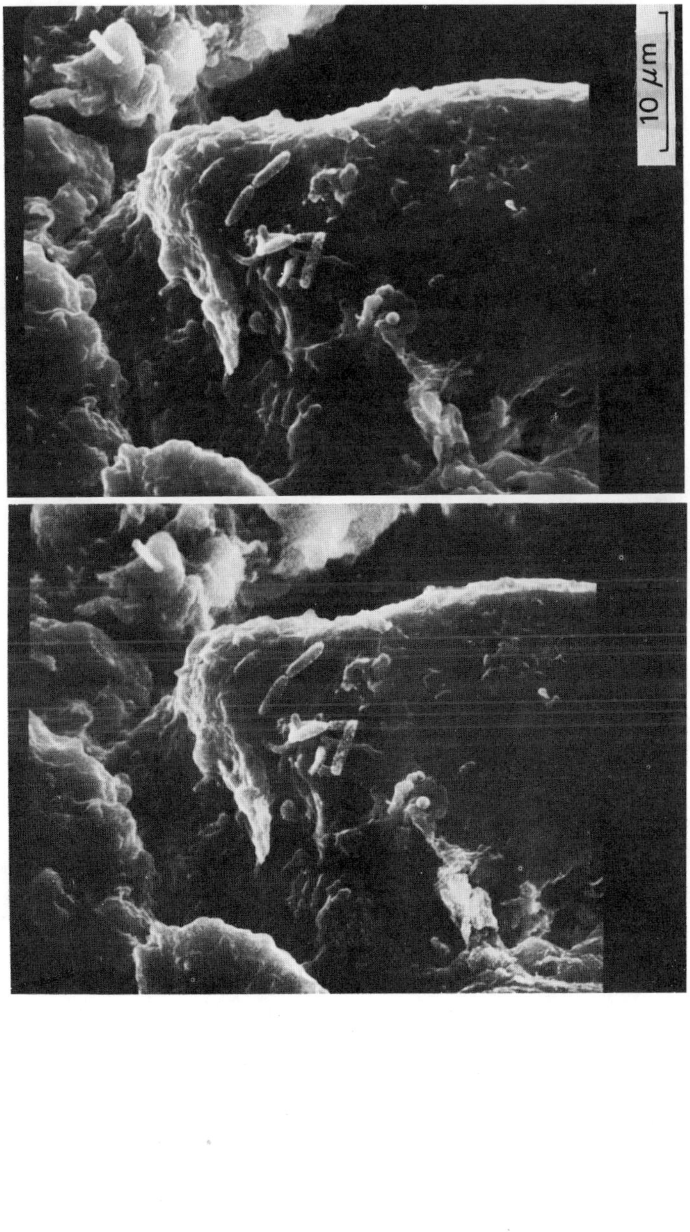

Fig. 9.10. Bacteria *in situ* in a red earth soil; stereoscopic SEM at 45/39°. The surfaces are probably clay cutans covering quartz grains. Supplied by K. C. Marshall. (From Marshall 1975.)

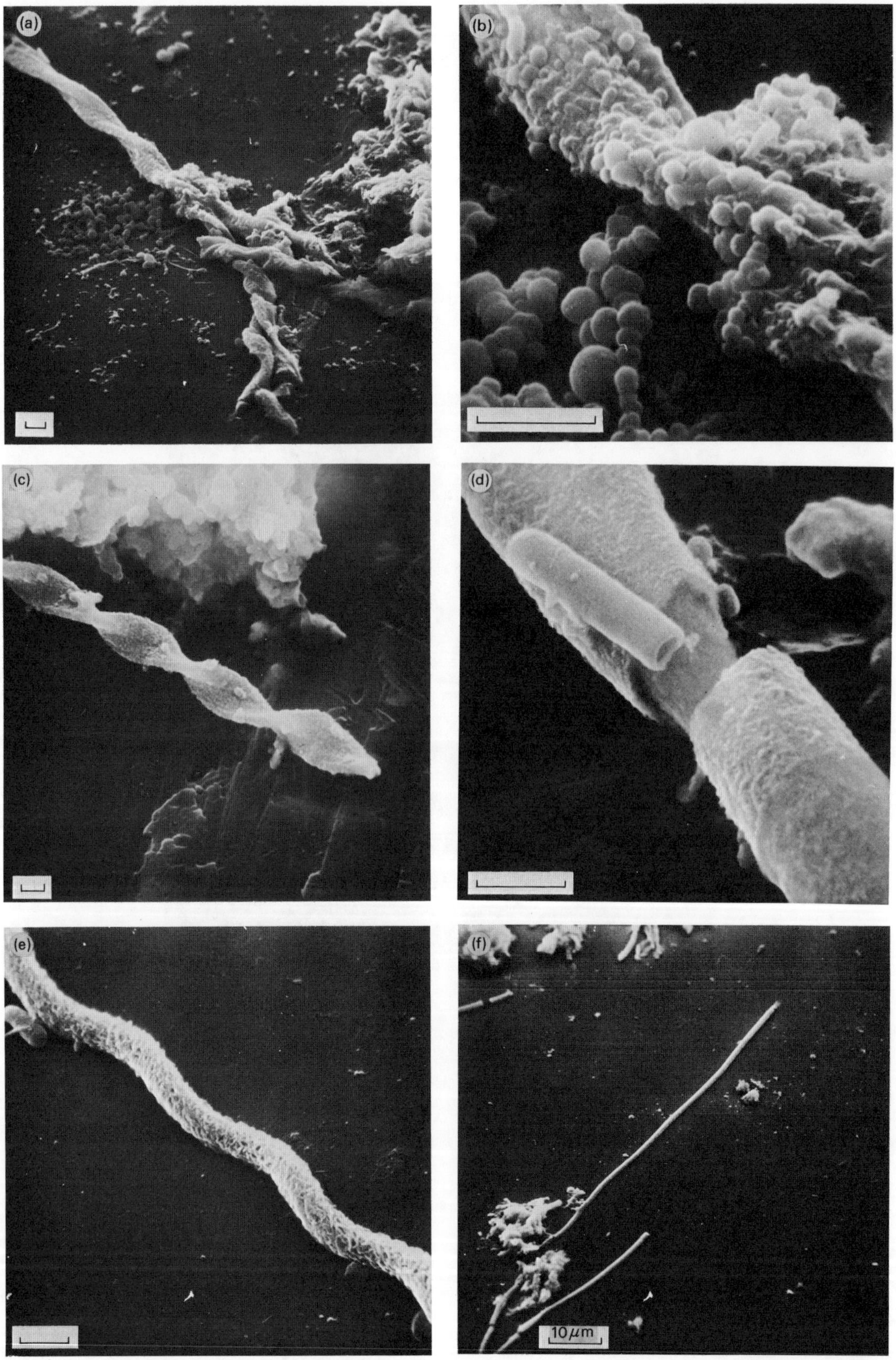
(a)
(b)
(c)
(d)
(e)
(f)
10 μm

9.12a

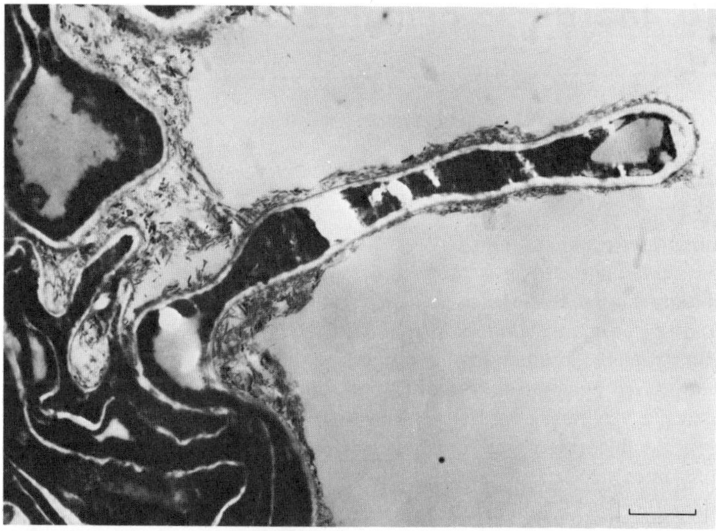

9.12b

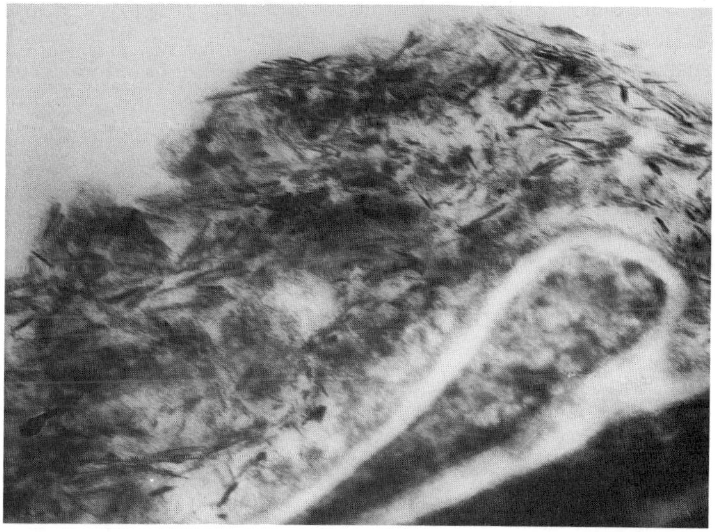

◁ **Fig. 9.11.** SEMs of filamentous ochre, such as fills agricultural field-drains, associated with various bacteria: (a) general view; (b) detail; (c) twisted ribbon of organic matter and iron hydroxide excreted by *Gallionella*; (d) *Leptothrix* in a gelatinous-hydroxide sheath, cigar-tube shaped item unidentified; (e–f) unknown. Supplied by C. Bloomfield. R. H. Turner, Rothamsted.

Fig. 9.12 Root hairs, (a) pea, (b) mustard; grown in γ-irradiated soil; roots selected; rinsed; cut to 1 mm length; glutaraldehyde; osmium tetroxide, or ruthenium red; ethanol, Durcapan ACM; ultrathin section; uranyl acetate, lead citrate; TEMs. (a) Some electron beam damage is visible as holes in the cytoplasm of the root hair, and there are a few holes in the section. There is a layer of mulcigel, which was exuded by the root, around the hair and beside the root proper. This contains many soil particles, which in this case are quite strongly oriented parallel to the surface of the root or hair. (b) Similar. $p = 1.4\,\mu m$. Supplied by J. F. Darbyshire. ((a) From Greaves and Darbyshire 1972; (b) from Darbyshire and Greaves 1973.) Macaulay Institute.

10 Inorganic cements

Important problems arise with sensitive clays. These have quite high strength until the stresses rise and initiate failure; then the strength is greatly reduced, and this occurs instantaneously, so that the failure is catastrophic. An alternative hypothesis is that failure is triggered by creep at constant stress. Landslides in sensitive clay have swept whole villages away with considerable loss of life. Some of these clays have a very open structure which is stabilized by a relatively weak inorganic cement; when the first few cementitious bonds are broken, the loads are thrown onto the others, and the whole structure of the soil collapses. The first examples here illustrate the studies which have been made of cementation of sensitive clays. The other examples are from laboratory studies of natural cementation and of soil stabilization, respectively.

Fig. 10.1. Leda Clay, Ottawa Sewage Works, depth 15 m (50 ft); undisturbed; alcohol-substitution dried; fracture surface; 30 nm Al; SEM, horizontal view of vertical fracture, θ 60°. This was tentatively interpreted as a bump, which looks like a quartz particle clothed in smaller particles, i.e. Kubiena's (1938) porphyropeptic fabric, which is associated with cementation. Sample obtained through C. B. Crawford. (From Smart, P. (1966). Soil structure, mechanical properties and electronmicroscopy. Ph.D. thesis, University of Cambridge.) H.190, M. Moss, Cambridge, 94.101.

Fig. 10.2. Leda Clay; undisturbed; air-dried; 20 nm Au, multidirectional; SEM, p 8 μm, θ 35°. Note the large range of particle sizes, the openness of the structure, and the way in which the smaller particles adhere to the larger. The openness is a prerequisite for sensitivity. The adherence suggests cementation, but not conclusively. Sample supplied by Queen's University, Ontario. (From Tovey, N. K. (1971). A selection of scanning electron micrographs of clays. Cambridge University Engineering Department Technical Report CUED/C—SOILS/TR5.) 375.35, Cambridge, UL.94.101.2A.2V.

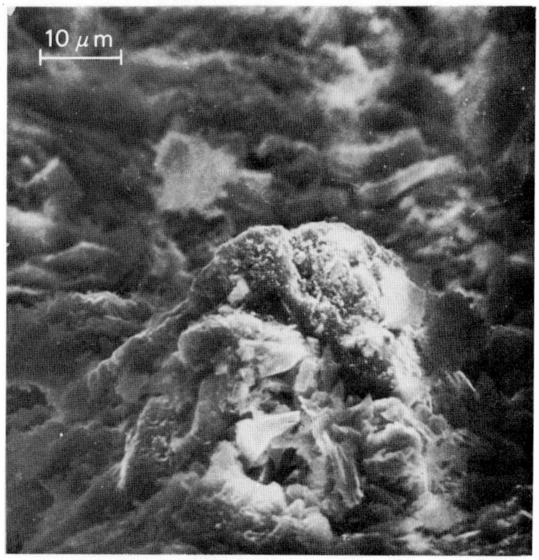

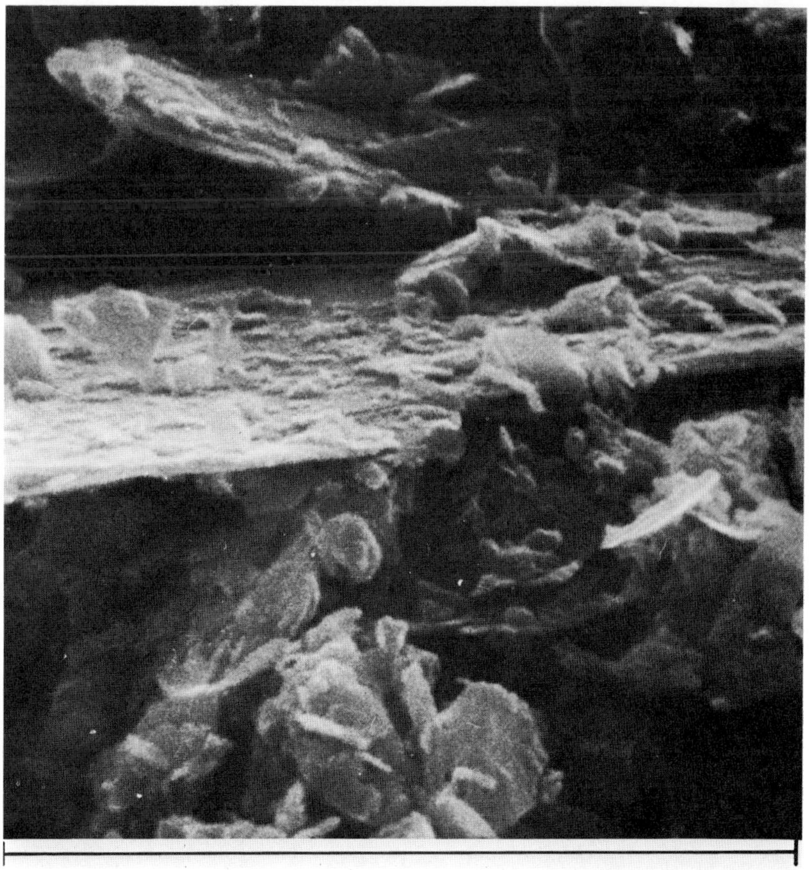

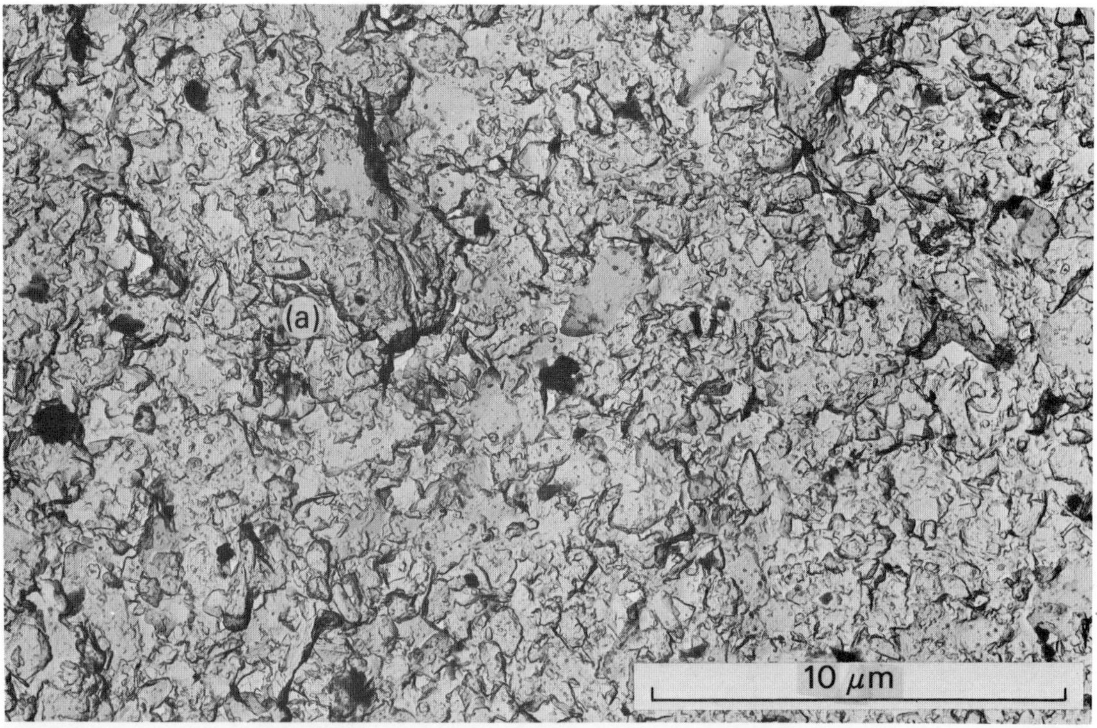
(a)
10 μm

(b)
10 μm

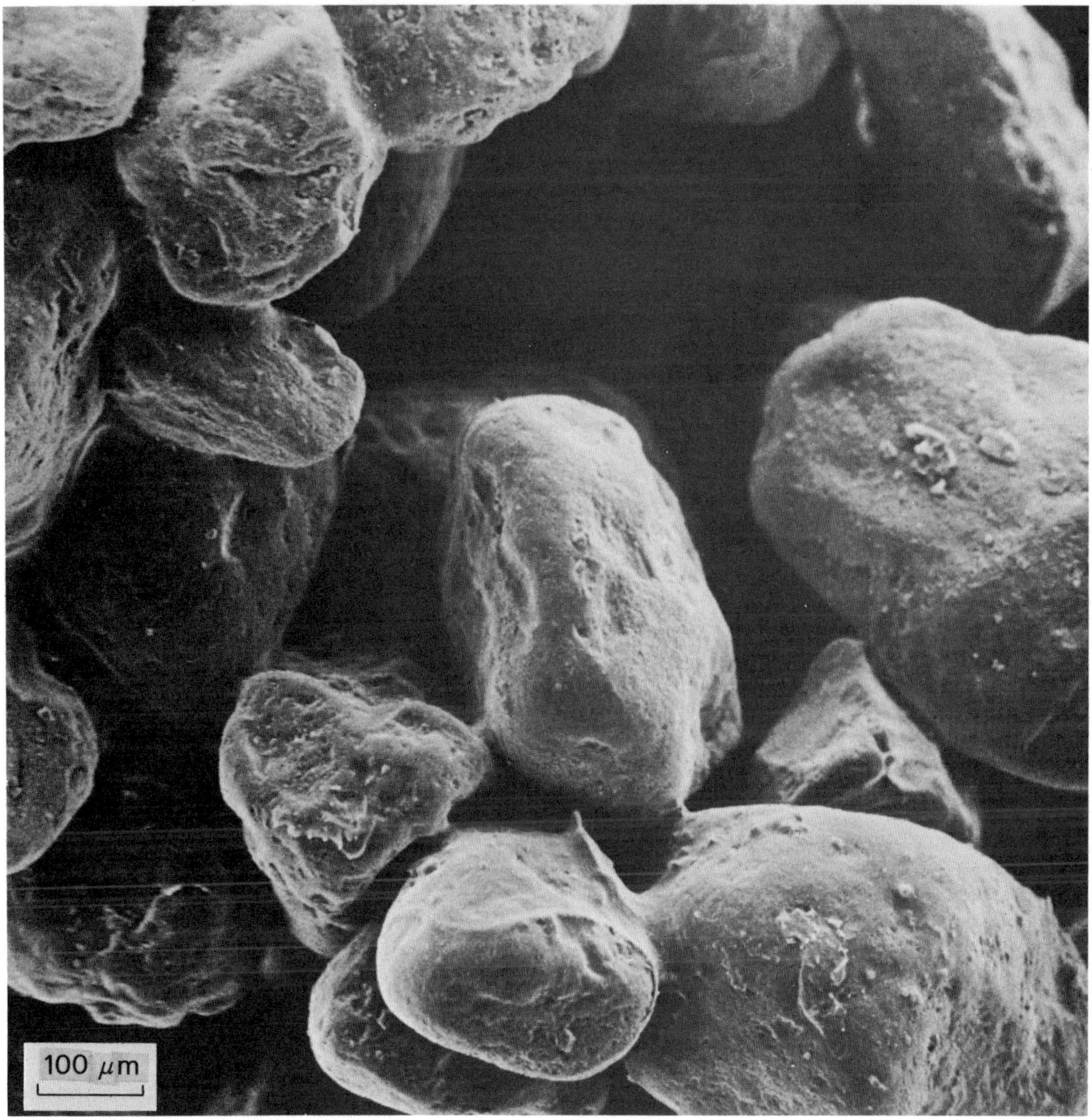

◁ **Fig. 10.3.** Champlain Sea Clay, St. Vallier. Quebec; (a) natural; (b) after selective dissolution of amorphous material using 8.0 N HCl and 0.5 N NaOH alternately and repeatedly; oriented slides; metal-shadowed carbon replicas; cleaned in concentrated HF and 0.2 N NaOH alternately; TEMs. The natural particles are coated with amorphous material, which other studies suggest to be silica-iron hydroxides, and which probably acted as cement. Supplied by R. N. Yong. (From McKyes, Sethi, and Yong 1974.) McGill University.

Fig. 10.4. Artificial silcrete, quartz sand grains cemented with silica; SEM. Little cement is yet present; it forms a thin layer over the two grains at the bottom and small bridges to the grain in the centre. Details; glacio-fluvial sand after 20 days in a ball mill. Supplied by W. B. Whalley. (From Whalley 1977.) The Queen's University of Belfast.

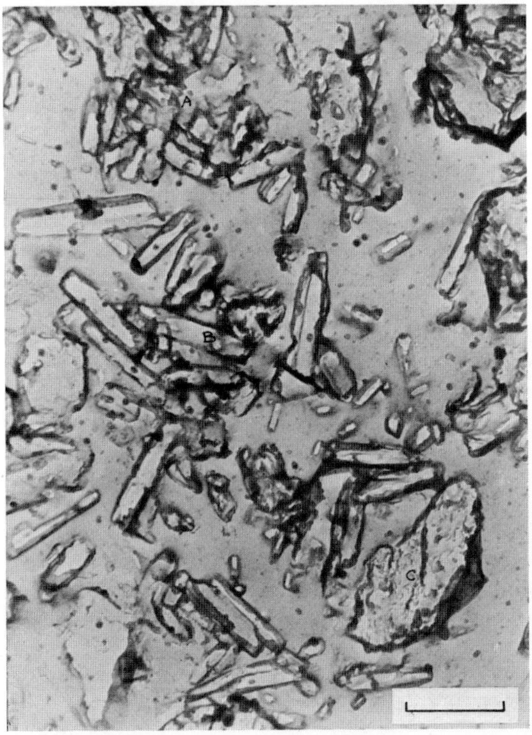

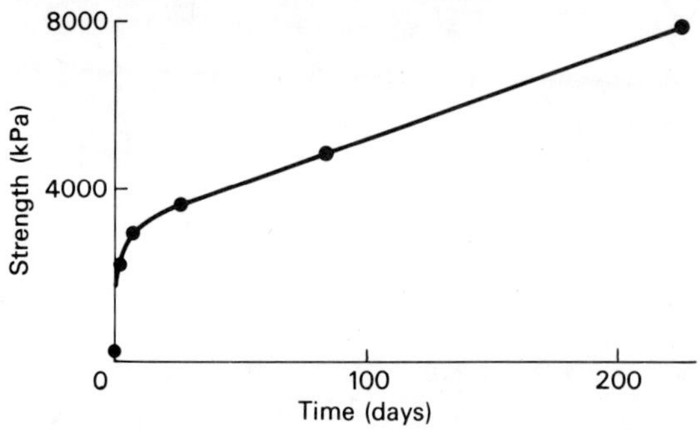

Fig. 10.5. Silica flour, montmorillonite, cement, and water; mixed; compacted; cured for 7 days; air-dried; fractured; replica; TEM. At this stage of the reaction, both the soil and the cement particles were being attacked and eaten away, often from all sides, forming an intermediate reaction product, mainly calcium silicate hydrate gel. By this stage, the strength had risen considerably, see graph. At a later stage, new particles formed from the gel, leading to a further increase in strength. Details: loose particles blown off fracture surface; pressed against polystyrene at 160 °C; sample removed in 48 per cent HF; rinsed in distilled water; air-dried; Pt-shadowed at 30–45°; carbon film; scored into 2 × 2 mm squares; polystyrene removed in ethylene dichloride. Supplied by J. K. Mitchell. (From Mitchell and El Jack 1965.) University of California at Berkeley.

11 Authigenic features

This and the next Section show authigenic and weathering features respectively. On the whole, we have allocated the examples to these two Sections according as the features are primarily being built up or broken down respectively. Thus, this Section shows some features formed *in situ*, mostly well developed ones, which probably grew freely into voids. Many of the clay minerals in Section 2 and the inter-grown domains in Section 6 were probably authigenic. Crystals, which grew in restricted spaces, and which were unable to expand these spaces, have shapes moulded by these spaces; we are not illustrating these. We therefore start by showing a crystal which may have been able to exert sufficient pressure to expand the void in which it was growing. Now, inter-particle structures are classed in two groups, according as they are formed by physical or chemical factors respectively (Berrier, personal communication). The previous Sections were more concerned with examples from the first of these groups; this Section shows examples from the second.† There are problems concerning the mechanisms of nucleation and growth of the crystals in these structures and the effects which the growth of these crystals in the voids have on the mechanical strength and transport properties of the soils concerned. Many of these features probably formed under saturated conditions. Towards the end of this Section, we include some concretions and cutans, some of which may have been formed under alternate wetting and drying conditions.

† The principal authigenic inter-particle structures are:

isolated fibres, rods, or tubes
individual fibres forming mat
individual fibres piled irregularly (probably sedimentary)
radiating bundles of rods
  sparse
  close packed
'mat grass'
inter-penetrating bundles of fibres
inter-woven sub-parallel bundles of fibres
parallel bundles of fibres
authigenic single-plate card-house
authigenic book-house
authigenic packet
globular cluster
  radial single plates
  radial multi-plate?
  random close-packed particles
  circumferential plates

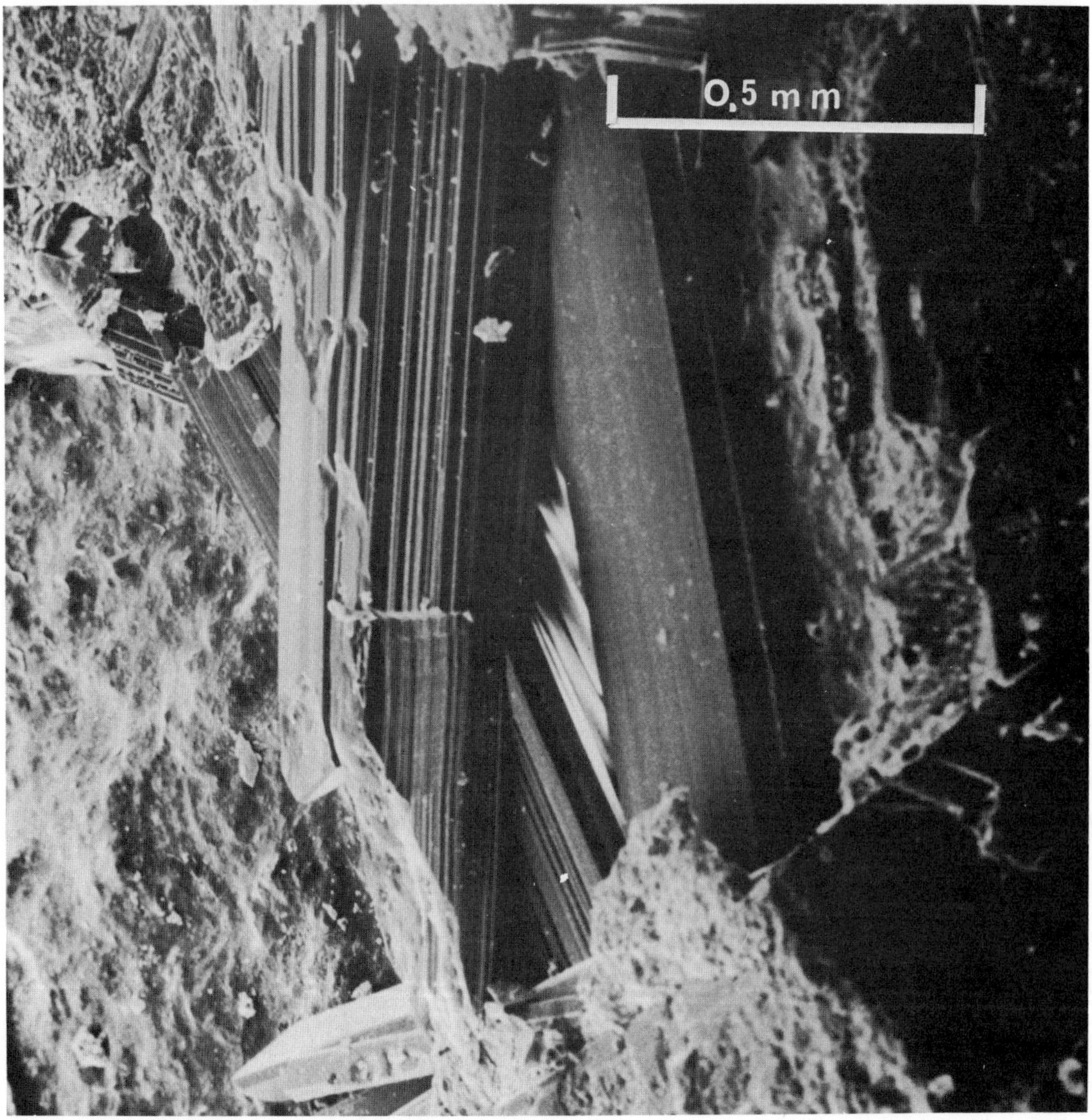

Fig. 11.1. Black, carbonaceous, pyritic shale, Billings Formation, Ordovician, from Ottawa; undisturbed sample; (fractured); 10 nm Au–Pd; SEM. This shows selenite (gypsum) crystals growing directly into the shale. Selenite crystals were thought to be opening fractures within the shale. Supplied by R. M. Quigley. (From Quigley, Zajic, McKeys, and Yong 1973.)

Fig. 11.2. Imogolite forming on plagioclase, weathered volcanic ash, San-in District, Japan; selected grains; washed; air-dried; (a) coated with C, then with Au; SEM; (b) cleavage surface; replica, shadowed at 45°; TEM. Note the white shadows on the replica. The imogolite starts to grow, presumably from solution, as isolated fibres 10–30 nm ϕ, perpendicular to the surface. Supplied by K. Tazaki. (From Tazaki 1979.) ▷

11.2a

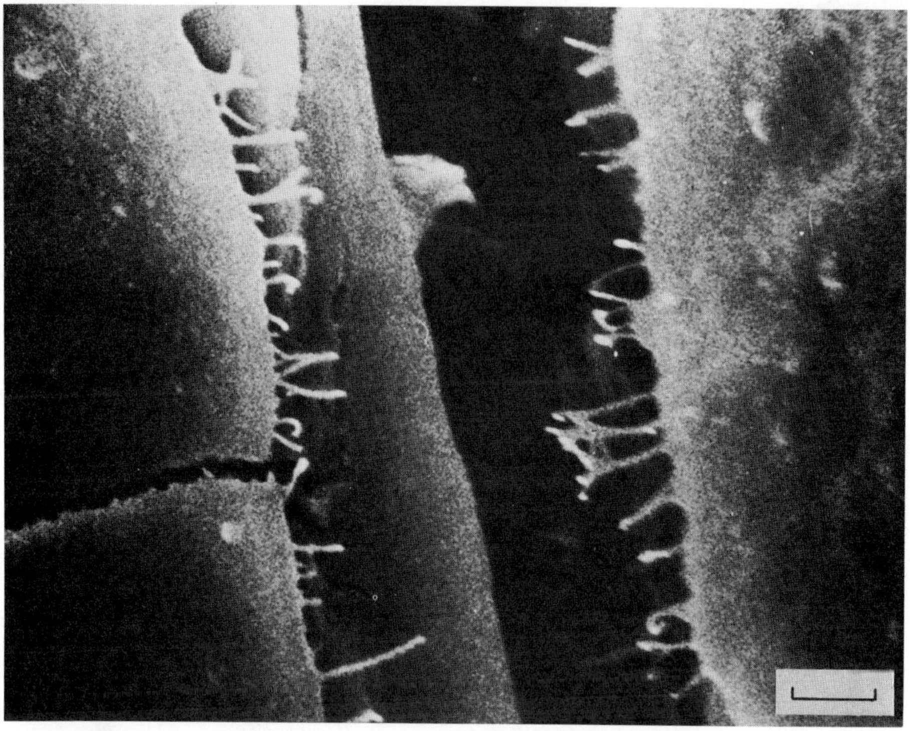

11.2b

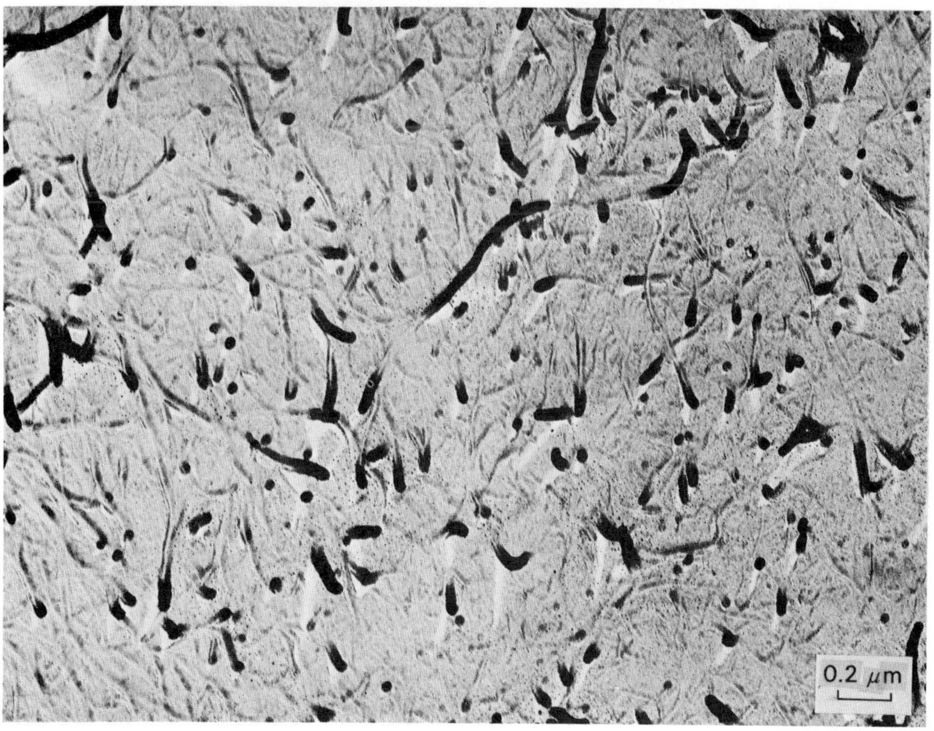

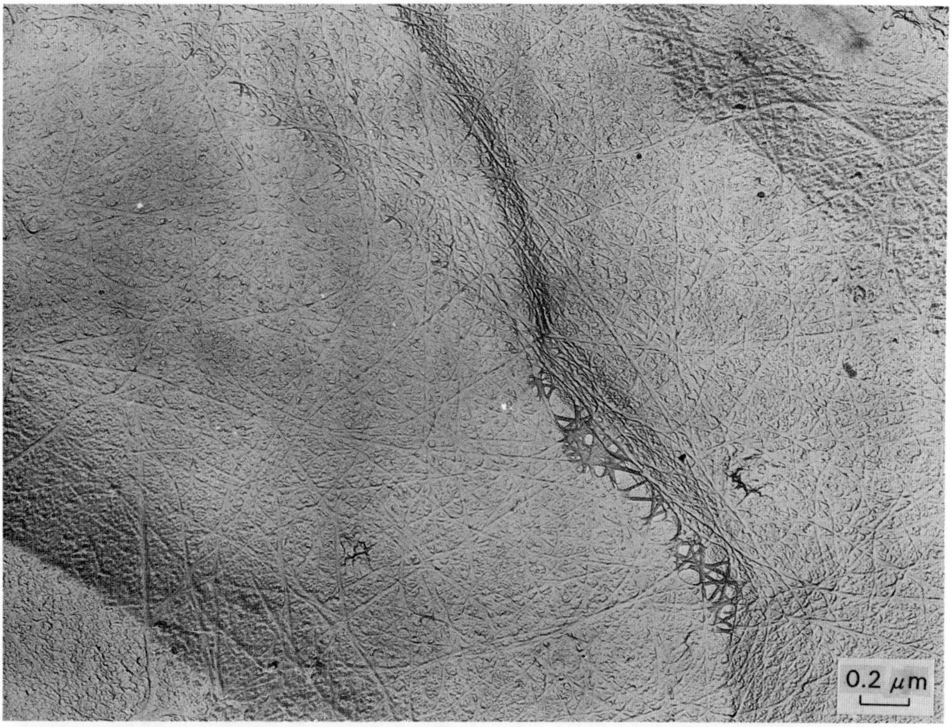

11.3

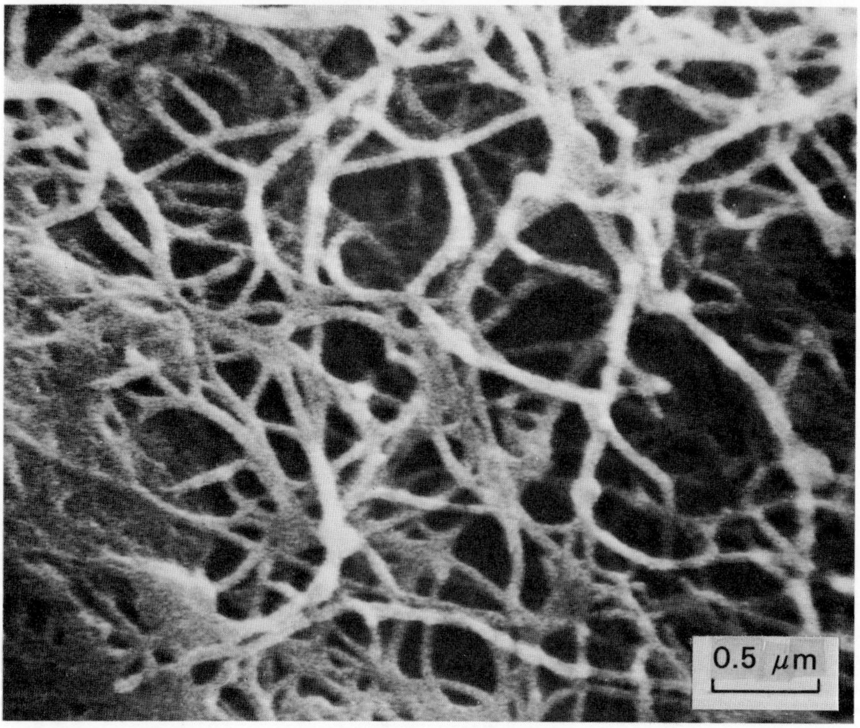

11.4

◁ **Fig. 11.3.** Same series as Fig. 11.2; replica. At a later stage, the imogolite threads form a mat covering the whole surface. Supplied by K. Tazaki. (From Tazaki 1979.)

Fig. 11.4. Imogolite from the Kurayoshi pumice beds; OsO_4 treated; ethanol, CO_2, supercritical-dried; coated with C, then with Au–Pd; SEM. Note the softly curved and entangled threads, 28–56 nm ϕ, which were better preserved by supercritical-drying than by air-drying. Supplied by K. Tazaki. (From Noishiki and Tazaki 1975.)

Fig. 11.5. Kaolin elongates (endellite) on the walls of a vug in perlite from Etzatlan, Mexico; SEM. This shows radiating bundles of rods with sparse space filling. This was placed by Keller (1978) in his general category of kaolin deposited from solution on an aluminium–silicate substrate. Supplied by W. D. Keller.

Fig. 11.6. Endellite–halloysite along microfractures(?) in kaolinizing Sparta Granite, Martin-Marietta Quarry, Sparta, Georgia; SEM. Included to illustrate 'mat grass', i.e. bundles of converging fibres, see the upper part of the micrograph. This material was categorized as kaolin deposited from solution on an aluminium–silicate substrate. Supplied by W. D. Keller. (From Keller 1978.)

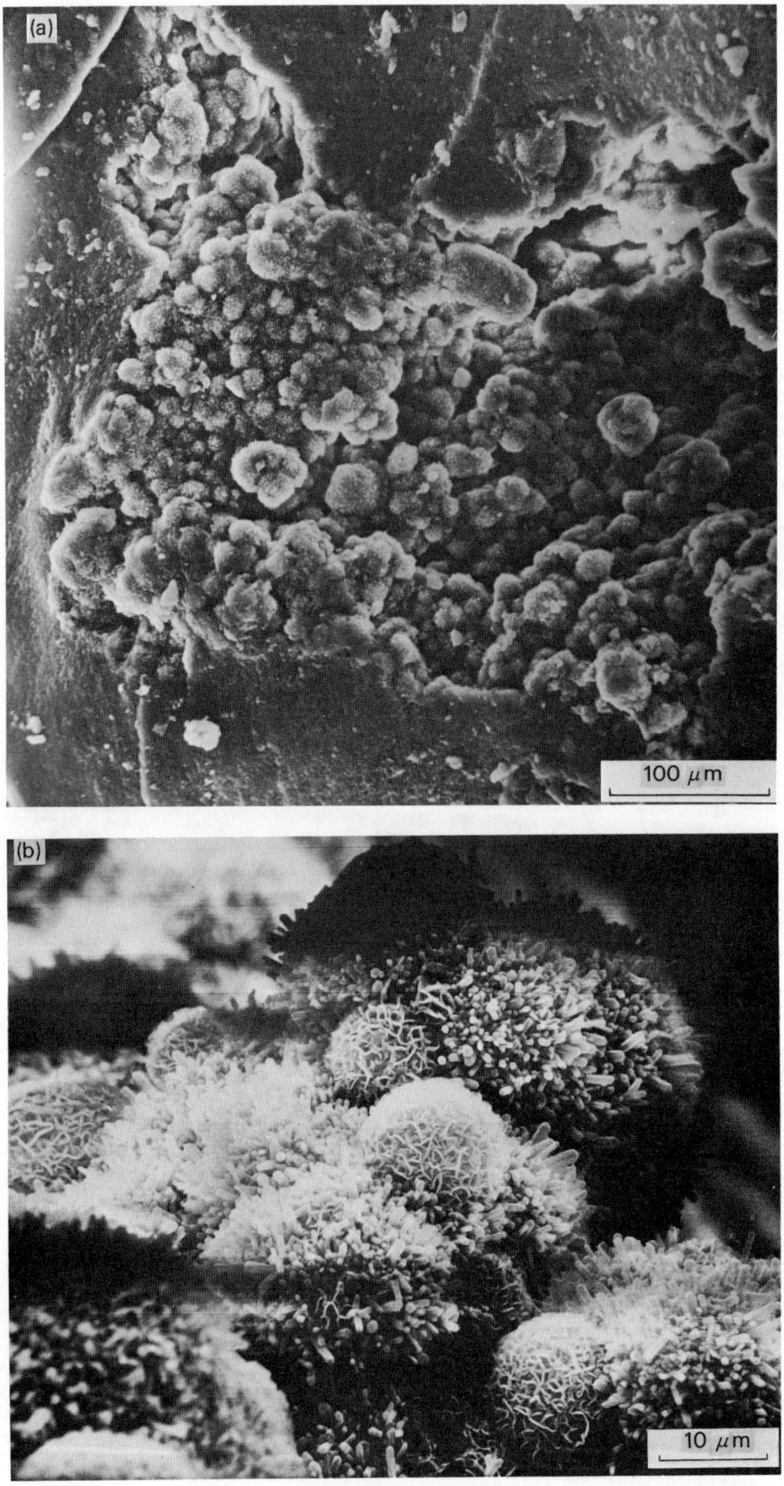
(a)
100 μm
(b)
10 μm

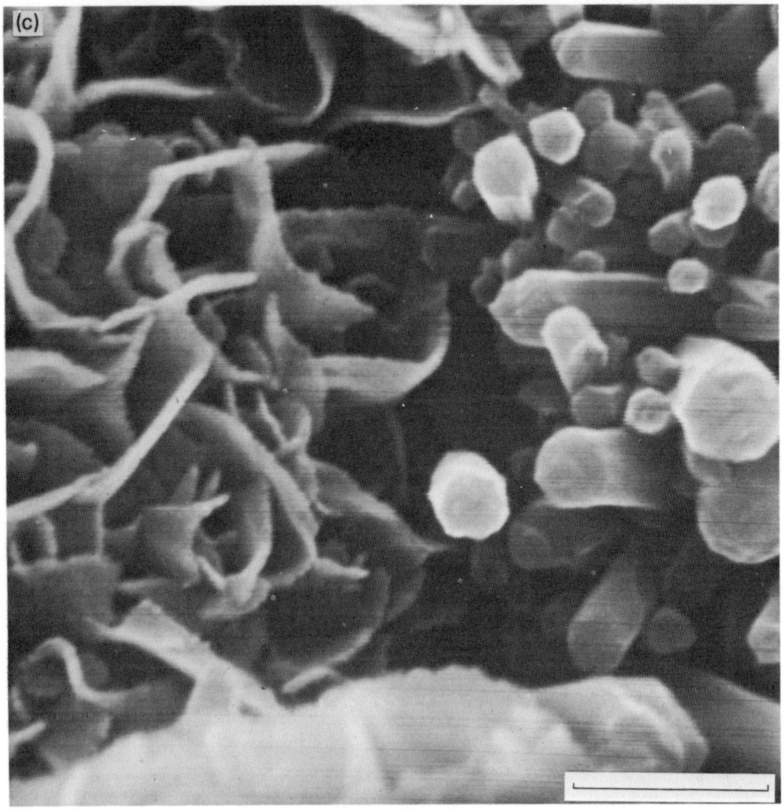

Fig. 11.7. Endellite (hydrated halloysite) from Gardner Mine Ridge, Bedford, Indiana; undisturbed sample; stored in sealed containers under water; (a) dissected to 5 mm cube and vacuum-dried (air-dried); (b) and (c) oven-dried at 105°C and fractured; All: 5 nm C, 25 nm Au–Pd; SEMs. These show globular clusters of radiating bundles of close-packed rods or tubes, also artefacts: (a) smearing during dissection; (b) and (c) unrolling of tubes during oven-drying. The clusters, 10–20 μm ϕ, are thought to have grown from solution. Supplied by S. Diamond. (From Diamond and Bloor 1970.)

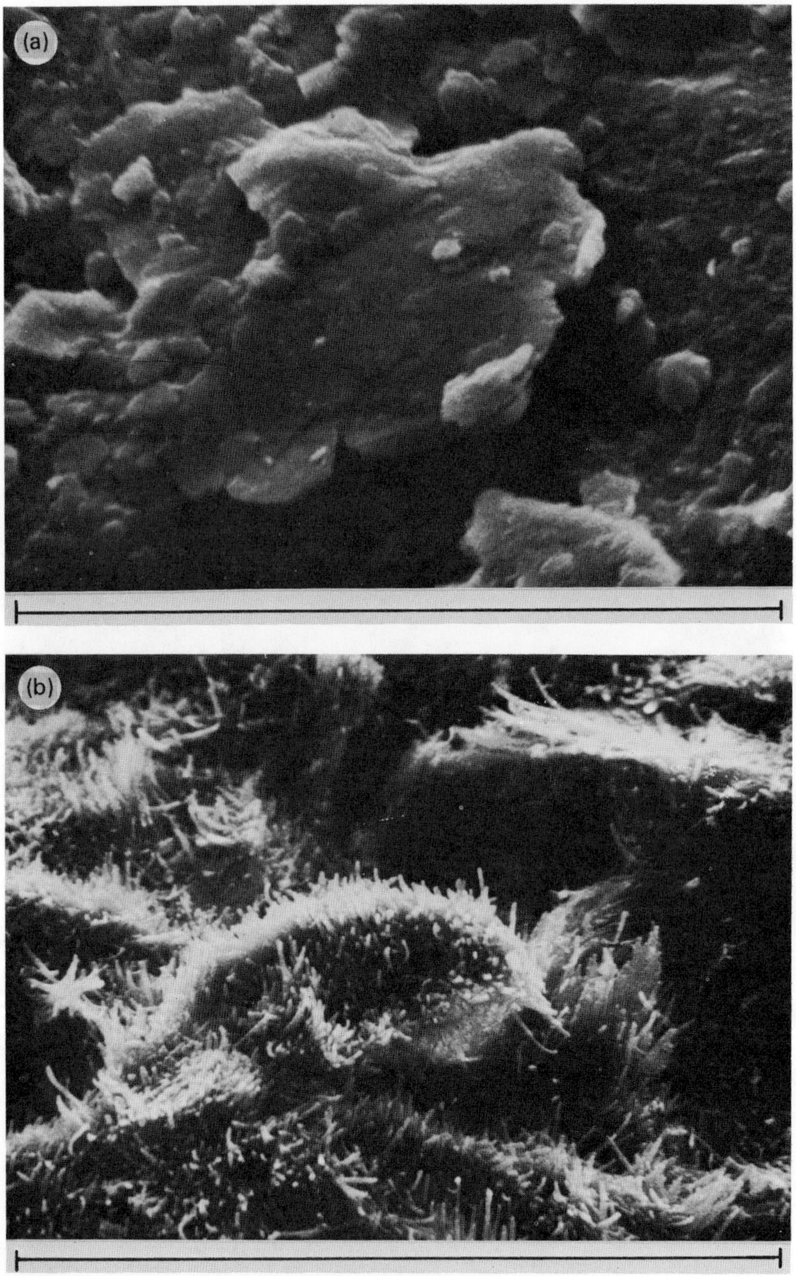

Fig. 11.8. SEMs showing alteration of lizardite to chrysotile. (a) Unaltered plates of lizardite; from Table Mountain, Newfoundland; *p* 18 μm. (b) Complete conversion by all round growth to chrysotile fibres; from Palmer Ridge; *p* 21 μm. (From Prichard 1979.) Taken and supplied by H. M. Prichard.

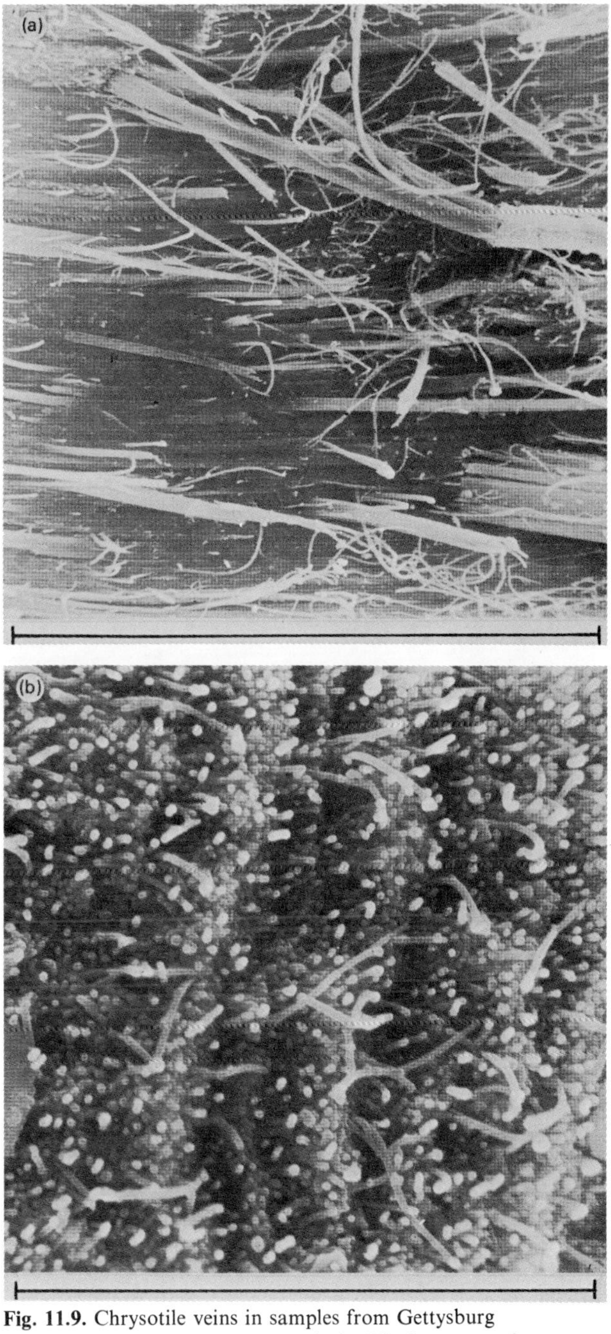

Fig. 11.9. Chrysotile veins in samples from Gettysburg Bank; SEMs, *p* 10, 2.5 μm respectively. The long axes of the fibres lie perpendicular to the axis of the vein in both cases. (a) The individual fibres extend across the whole width of the vein. (b) Short fibres lie in rows giving a segmented appearance. (From Prichard 1979.) Taken and supplied by H. M. Prichard, University of Reading.

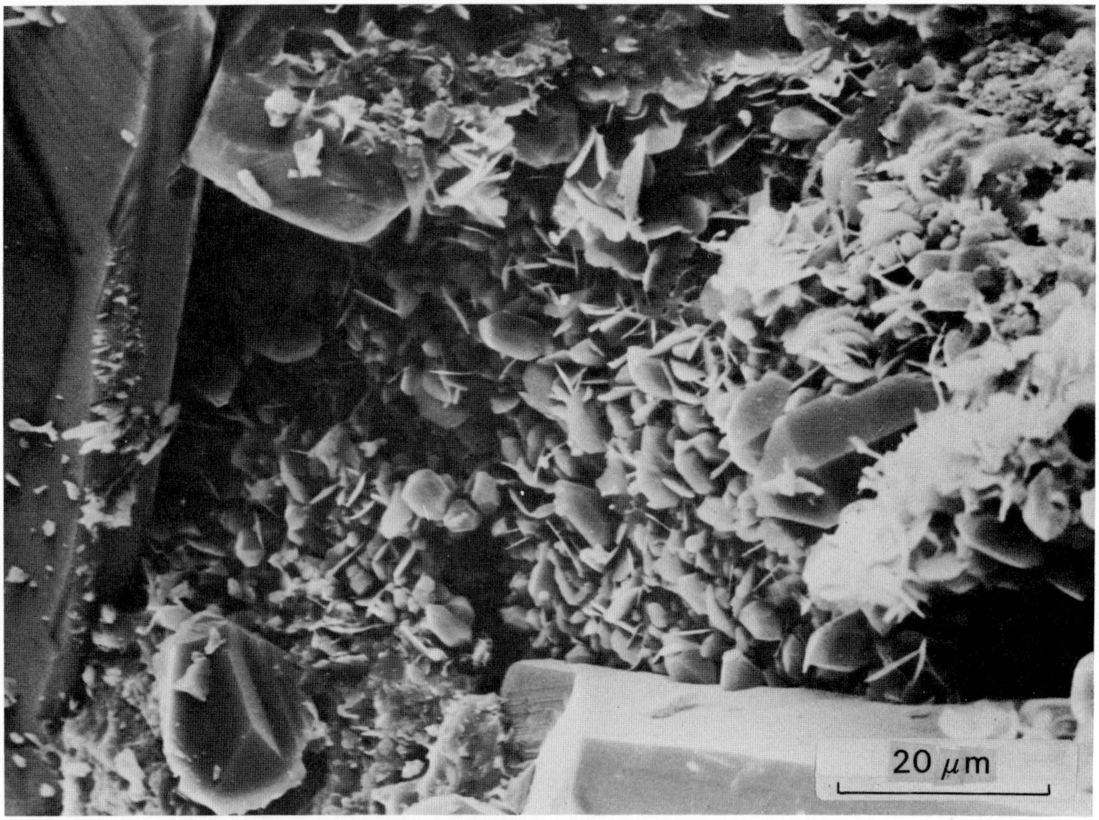

Fig. 11.10. Chlorite, Fahler Formation, Cretaceous, from N.W. Alberta; undisturbed sample; fracture surface; SEM. This is best described as authigenic single-plate card-house structure. Micrograph supplied by J. E. Gillott (shown at a meeting of the Royal Microscopical Society, Edinburgh, 1979). Neg. 00 048 S.

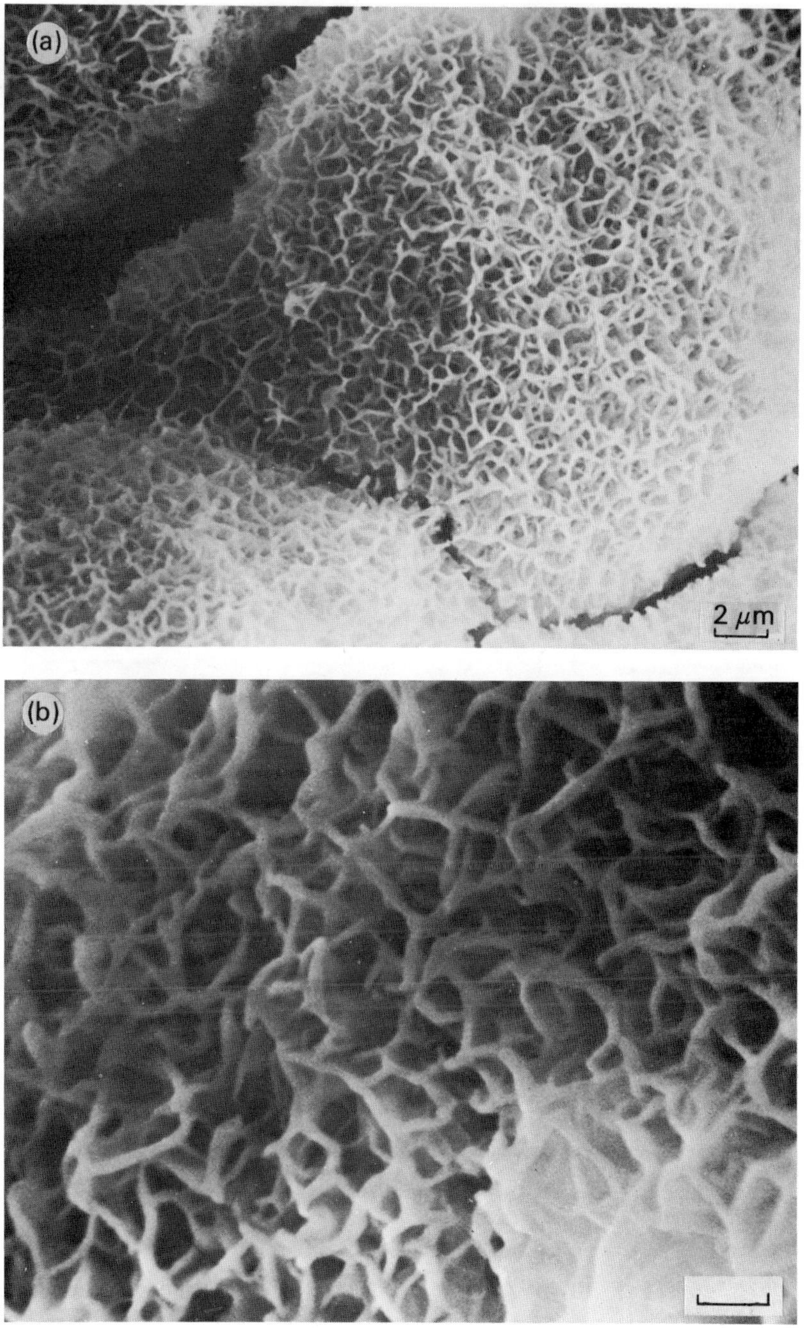

Fig. 11.11. Montmorillonite in lithic tuff in submarine sediment from Mariana Trench, Pacific Ocean, under 5450 m water; SEMs. (b) is a detail of (a). This might almost be described as a globular cluster with radial single plates, i.e. with single-plate card-house structure; but the structural units could contain several layers of particles, in which case, the fine structure might be better described as tactoid structure. Supplied by K. Tazaki. (From Tazaki, Tazaki, and Komatsu 1978.) Both from sample ST.1404.D1.3.2.

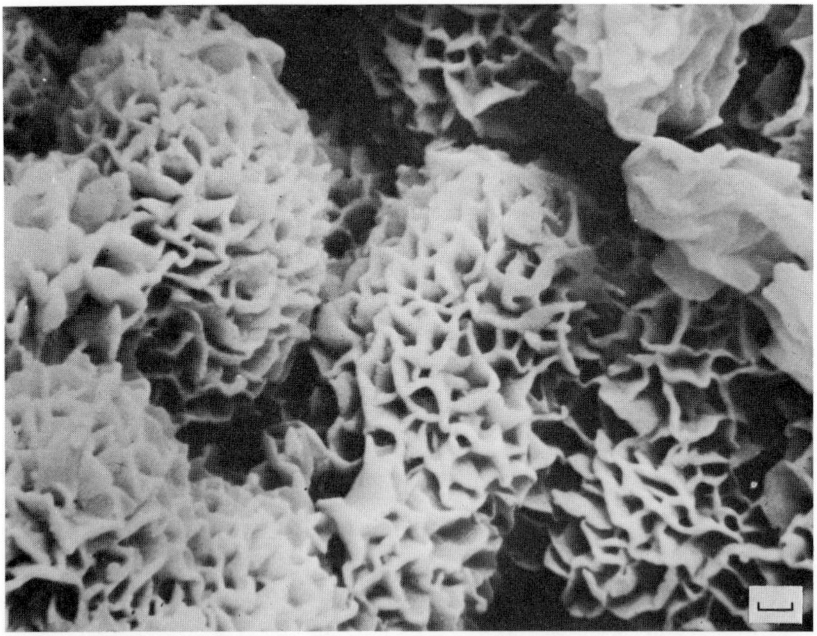

11.12

11.13

Fig. 11.12. Nontronite in eluvial clay from near Sverdlovsk, 4 m depth; undisturbed sample; freeze-dried; ground on emery paper, peeled with adhesive tape; SEM. Comments as preceding figure. (From Sergeyev, Grabowska-Olszewska, Osipov, and Sokolov 1978; Osipov and Sokolov 1978.) Moscow State University.

Fig. 11.13. Montmorillonite in eluvial clay from Chmielnik; undisturbed sample; freeze-dried peeled; SEM. Globular clusters, perhaps slightly degenerate, containing somewhat close-packed plates or flakes. (From Sergeyev *et al.* 1978.) Moscow State University.

11.14

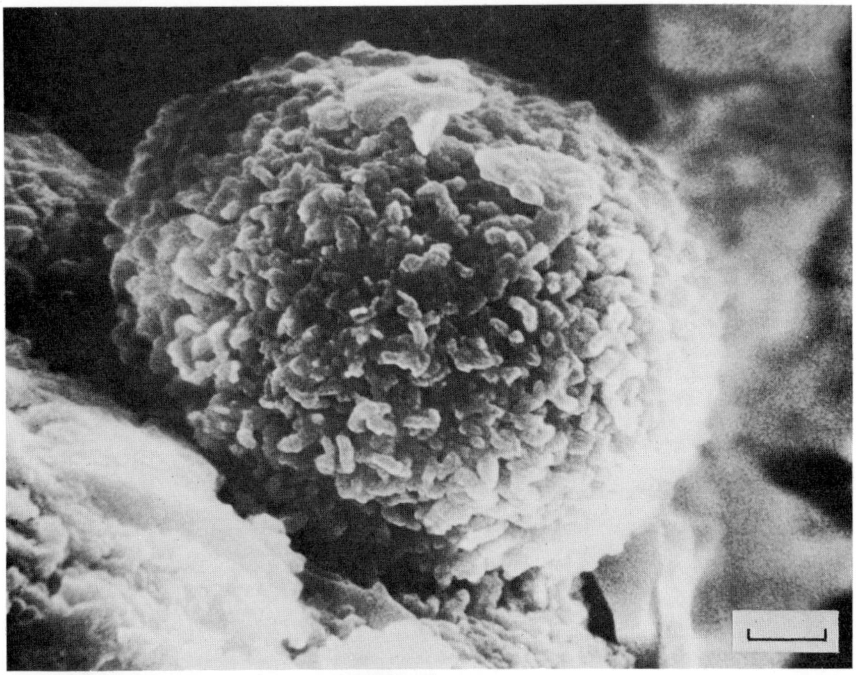

Fig. 11.14. Halloysite forming from plagioclase; SEM. Tentatively a globular cluster with close-packed random structure. Tentatively, Tazaki's class Ht 2–4. Supplied by K. Tazaki. (From Tazaki 1978*a*, *b*.)

Fig. 11.15. Carbonate nodule; (a) exterior; (b) fracture surface showing interior of nodule; SEMs. (a) The nodule is composed of microcalcites up to 4 μm long. (b) The microcalcites are in turn composed of oval particles, 0.2–0.6 μm ϕ. The forms taken are probably controlled by the clay minerals present and/or soluble organic matter. Supplied by D. H. Yaalon. ((b) From Wieder and Yaalon 1974.) Hebrew University, Jerusalem.

Fig. 11.16 (a). Pyrite in Drammen Clay; air-dried; 5 nm C, 20 nm Au; stereoscopic SEMs, *p* 9 μm, *θ* 40/35°. The individual crystals are octahedral in shape and randomly aggregated into a nodule. Sample supplied by A. N. Schofield. 316.8 & 316.9, Cambridge.

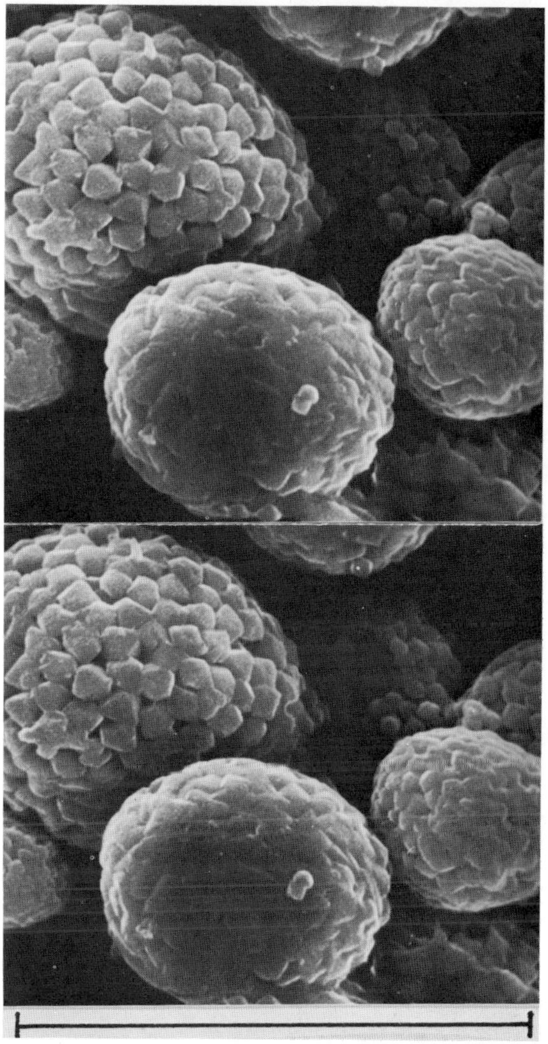

Fig. 11.16(b). Pyrite in Littleport (Burnt Fen) Clay; air-dried; 5 nm C, 20 nm Au; stereoscopic SEMs, *p* 13 μm, *θ* 35/40°. The crystals are aggregated into nodules, some of which are covered with a coating; these nodules filled an area 5 nm *ϕ*. 232.36 & 232.37, Cambridge.

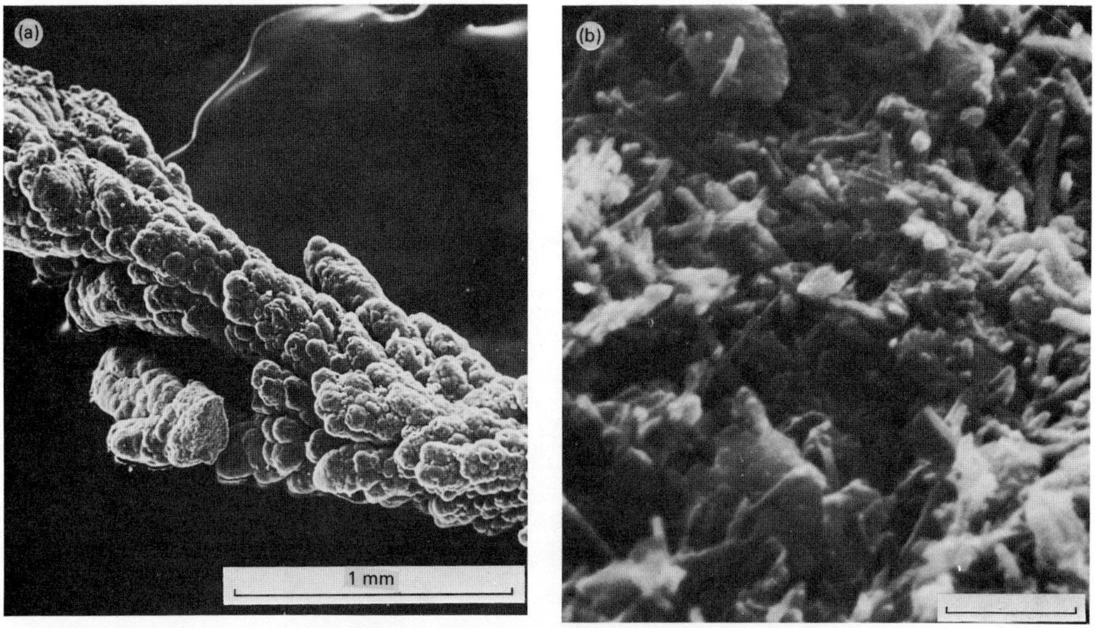

Fig. 11.17. Manganese oxide concretion from a pan, 2 m deep, at Birness, Aberdeenshire; SEMs. These concretions had a dull ochreous colour and fibrous texture. (a) Higher magnification shows a close-packed random mixture of plates(?) and laths. (b) The laths are hollandite or cryptomelane. Supplied by W. J. McHardy. (From McHardy 1977.) 15749 & 15745, W. J. McH., Macaulay Institute.

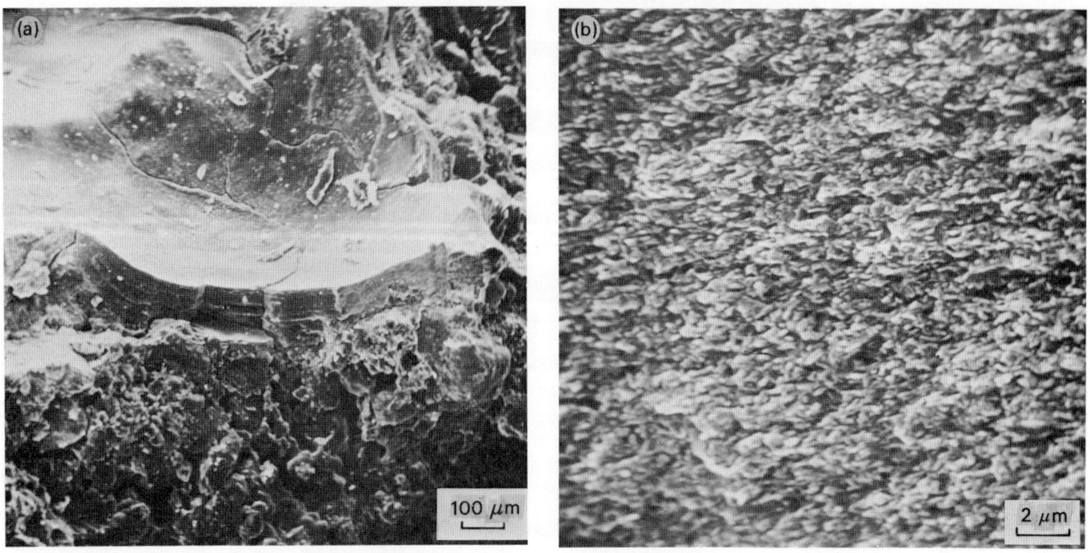

Fig. 11.18. Cutan, clay coating, from a pore in a duripan, Red Hardpan Soil, Mt. Keith area, Australia, depth 0.5 m; air-dried; Au; SEMs. (a) General view showing the face and edge of the coating and the underlying material. (b) Detail of the edge of the coating showing strongly oriented clay plates. Supplied by A. Low and E. A. FitzPatrick. 27366 & 28193, Macaulay Institute.

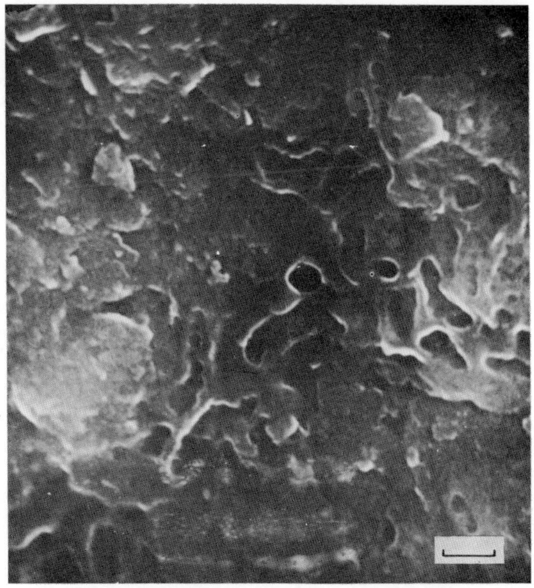

Fig. 11.19. Cutan, clay coating, from the floor of a pore, Bt horizon, Hamble Series, England, depth 0.45 m; air-dried; Au; SEM showing the surface of the coating. This pattern, with quite an open structure and uneven edges, similar to those of jigsaw pieces, seems to occur in several soil types (Low, personal communication). Supplied by A. Low and E. A. FitzPatrick. 26649, Macaulay Institute.

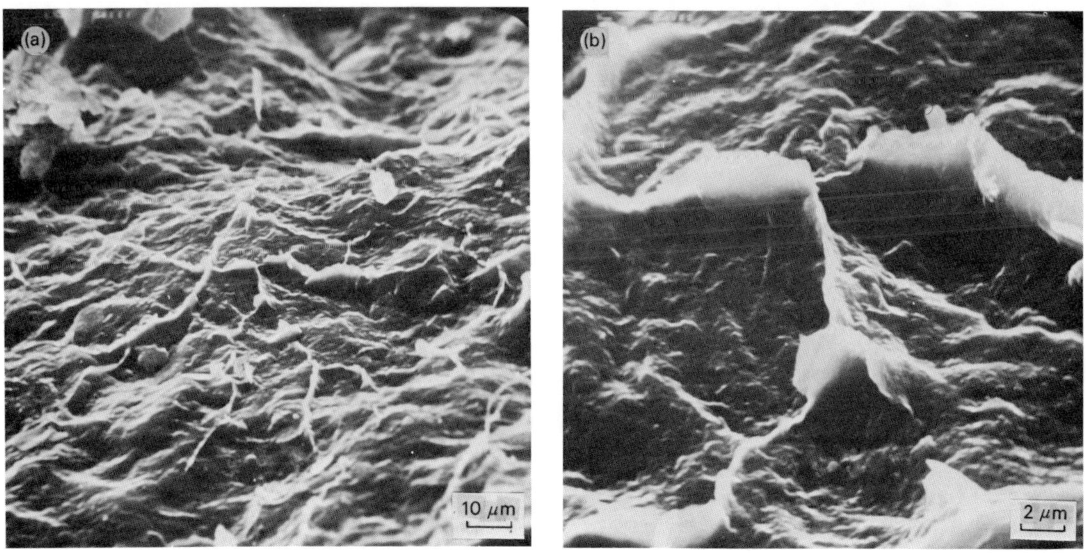

Fig. 11.20. Cutan, clay coating, from a pore, Iwo Series, Nigeria, depth 1.05 m, in tropical red and cream mottled clay formed from highly weathered granitic gneiss; air-dried, Au; SEMs of the surface of the coating. Note the ridges. Supplied by A. Low and E. A. FitzPatrick. 26697 and 26696, Macaulay Institute.

Fig. 11.21. Keuper Marl; ped face(?); air-dried; Pt-C shadowed at 30°, carbon backed replica; TEM; positive print, shadows white. The orientation of the plates and laths may have originated from shearing; but their indistinctness suggested recrystallization. Sample supplied by Road Research Laboratory. J.143.G, P.S., Nottingham, sample F.

Fig. 11.22. Waco Clay subsoil, Queensland, Australia; slickensided ped face; air-dried; Au, C; SEM, θ 45°. The slickensides were inclined at about 45° and resulted from alternate shrinking and swelling. The striations running left-to-right are mainly caused by differential movements, the marks inclined at about 80° to these striations include several root imprints, but there might have been movement in this direction also. 5355, A. Ismail, Glasgow, 360.S3. Sample located by G. G. Beckman and B. Bridge.

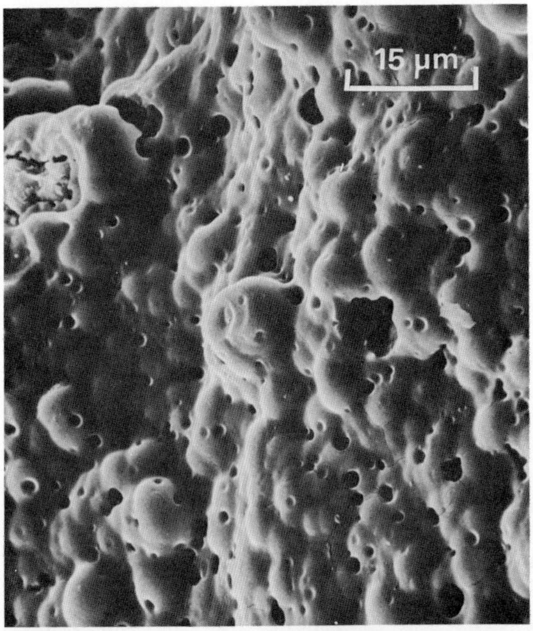

Fig. 11.23. Black, carbonaceous, pyritic shale, Billings Formation, Ordovician, from Ottawa; undisturbed sample; 10 nm Au–Pd; SEM. This shows a yellow alteration product, presumably jarosite, formed on the face of an open void under drying conditions. The material was probably non-crystalline. The holes, which are about 1 µm ϕ, may have been caused by bacteria, or the material may have been precipitated from water which oozed through these holes from the shale behind and then evaporated. Supplied by R. M. Quigley. (From Quigley *et al.* 1973.)

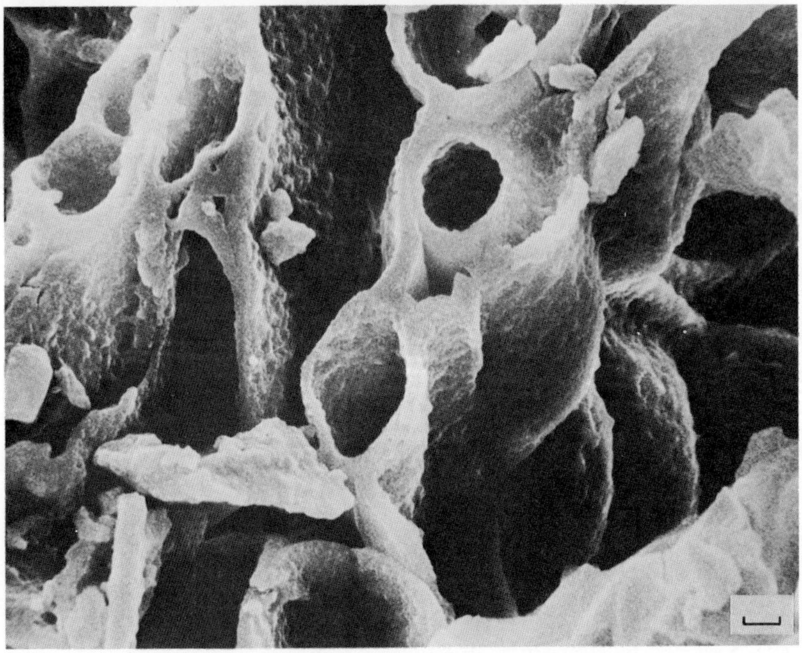

Fig. 11.24. Volcanic ash at incipient stage of weathering; SEM. The holes are perhaps due to the release of volatile gas from the underlying material. Supplied by K. Tazaki. (From Tazaki 1978*a*.)

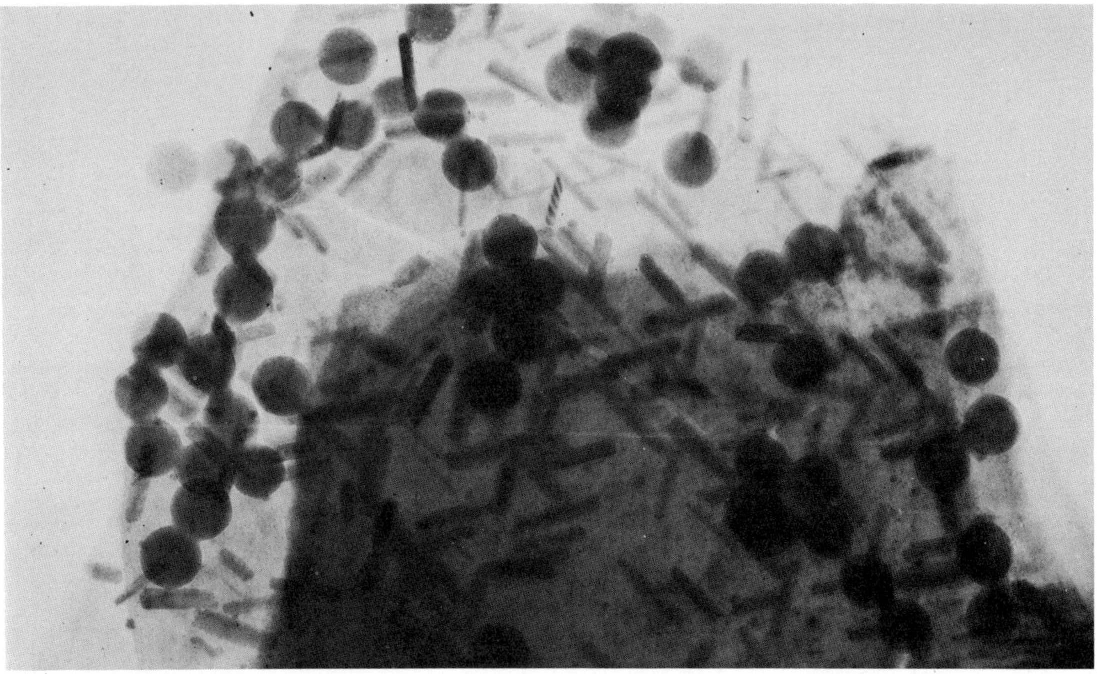

Fig. 11.25. Laths of β-FeOOH growing epitaxially on a vermiculite flake; many are aligned in the three hexagonal directions. Polystyrene spheres, 88 nm ϕ, have been added to check the magnification. Details: fresh green biotite; K replaced by Ba using 0.05 M $BaCl_2$; Ba replaced by Na using 1 M NaCl; oxidized in bromine water; Na-saturated using 1 M NaCl; ultrasonic dispersion; TEM. Supplied by W. J. McHardy. (From Farmer, Russell, McHardy, Newman, Ahlrichs, and Rimsaite 1971.) 67.747, W. J. McHardy, Macaulay Institute.

12 Weathering

The first part of this Section illustrates the disruption of soil structure by primarily physical effects; the second part illustrates chemical weathering of particles; and the last micrographs illustrate the alteration of soil structure by chemical effects. Some readers may find it helpful to have our notes on the weathering of particles, and we append these;† but our comments are rather tentative as further work seems needed.

†Weathering of Particles

1. Rounding of Corners
   Dissolution, without crystallographic control, most exposed atoms go first, initial stage of 4.1.1.
2. Edge attack
   - 2.1. Irregular edge attack
     Dislocations and impurities.
   - 2.2. Splitting along layers
     Planar structure with lattice strain
   - 2.3. Selective dissolution of certain layers
     e.g. 'inter-layer' from chlorite, or eutectics.
3. Face attack
   - 3.1. Etch pits
     Dislocations and impurities
   - 3.2. Transformation via fluid phase
     New and old crystals are face-to-face say 1 nm apart, possibly no visible symptoms once the crytals are separated except perhaps regular or irregular growth and dissolution steps.
   - 3.3. Transformation via intermediate solid phase
4. All round attack
   - 4.1. Transformation via fluid phase
     - 4.1.1. Without crystallographic control
       Old particles rounded, see 1 above, possibly of theoretical interest only.
     - 4.1.2. With crystallographic control
       Old particles roughly angular, with large voids between.
   - 4.2. Transformation via intermediate solid phase
   - 4.3. Solid state transformation
     - 4.3.1. Diffuse boundary
       Diffusion of atoms through crystal, small difference between new and old crystals.
     - 4.3.2. Sharp boundary
       Diffusion of atoms through crystal, relatively large difference between new and old crystals.

Fig. 12.1. Arid Red Earth from Alice Springs, Australia, depths: (a) ≈ 37 mm (1.5 in), (b) 1.65 m (65 in); dried; impregnated with methacrylate; ultrathin section; TEM. These are end members of a series showing progressive change from oriented domains with planar voids at depth to random structure near the surface. (From Silva, S. R., Spiers, V. M., and Gross, K. A. (1965). Soil fabric study with the electron microscope, a progress report. Defence Standards Laboratories Report 282.) Australian Defence Materials Research Laboratories.

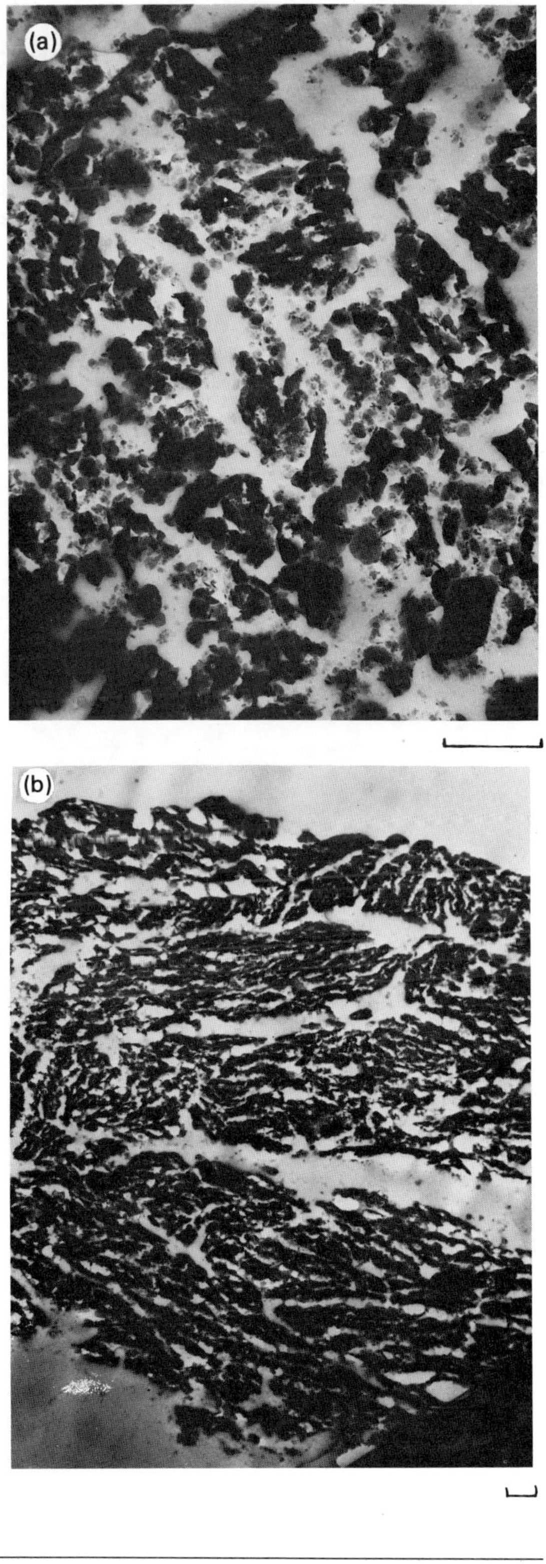

(a)

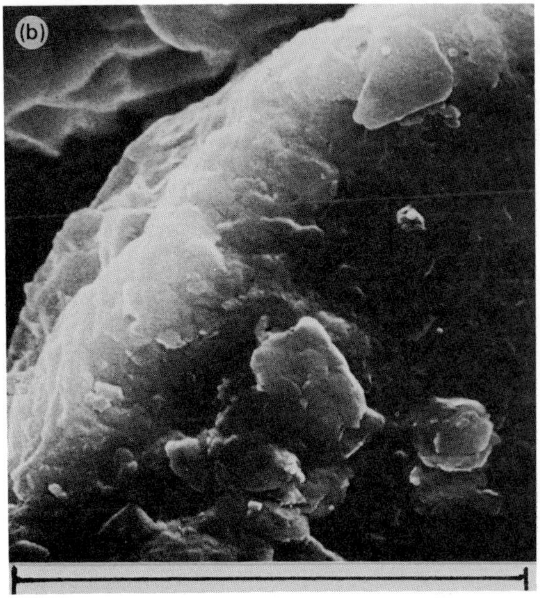

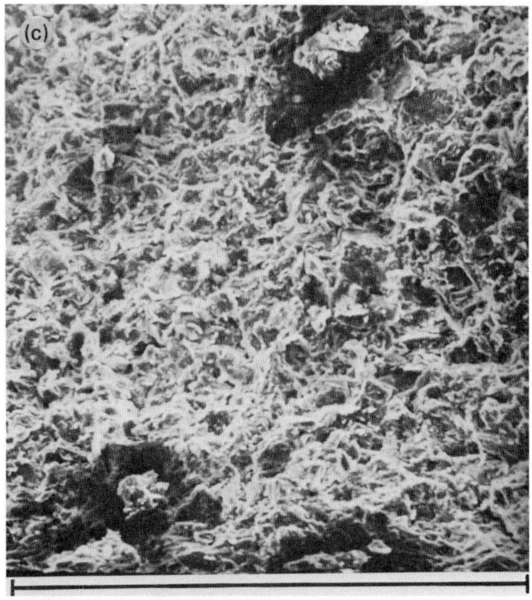

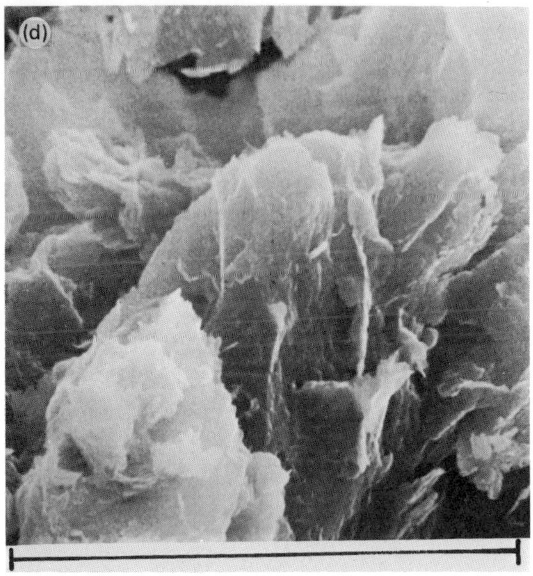

Fig. 12.2. Clayey agricultural topsoil; remoulded, compacted; (a) and (b) several freezing-thawing-drying wetting cycles. (c) and (d) Several drying-wetting cycles; air-dried; vertical fractures; 5 nm C, 20 nm Au; SEMs, *p* 700, 17, 700, 17 μm respectively, *θ* 42, 33, 42, 42° respectively. The freezing cycles, simulating an English winter, caused peds 30–150 μm *ϕ* and large cracks to develop, compare (a) with (c); the peds consisted of close-packed face-to-face aggregations with little evidence of intra-pedal voids (b), whereas the particles in the unfrozen samples formed a more open structure (d). The frozen samples had a much higher permeability, attributed to the cracks. Samples supplied by NAAS, Cambridge. 803.9A, 804.18A, 815.9A, and 815.6A, Cambridge.

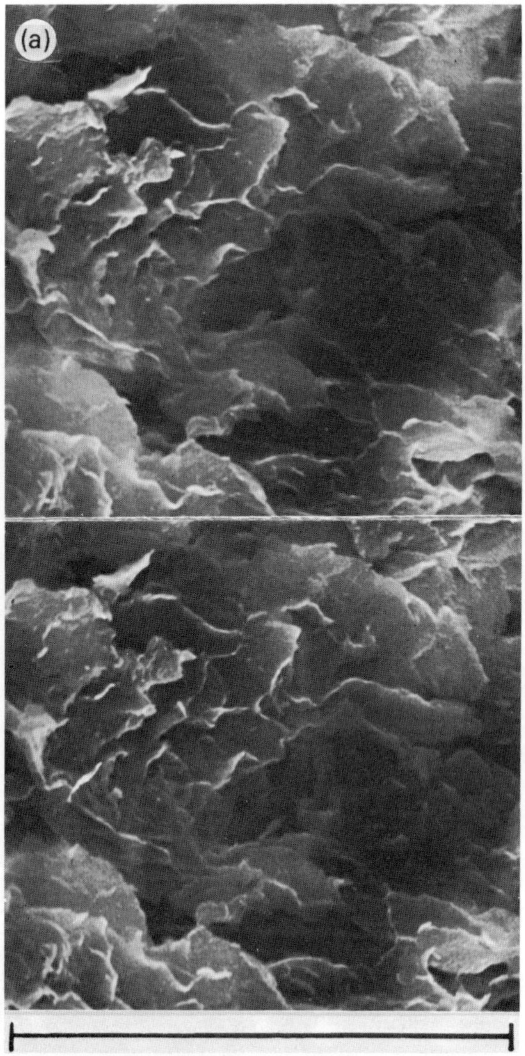

Fig. 12.3(a). Lud Clay, Israel; undisturbed sample; freeze-dried; vertical fracture surface; 5 nm C, 20 nm Au; stereoscopic SEMs, p 18.6 μm, θ 42/33°. The structure is characteristic of a clay with a high smectite content. Details: from 1.3 m depth below a covered area; w = 24.3 per cent S_r = 93 per cent; LL = 68 per cent; PL = 26 per cent. (From Tovey, Frydman, and Wong 1973.) 1115.14/15, Cambridge.

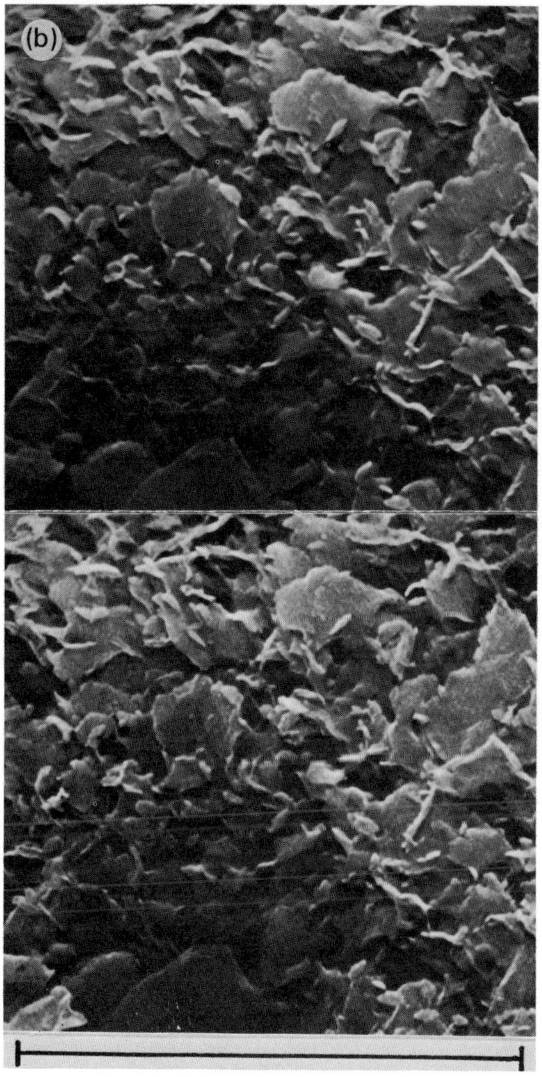

Fig. 12.3(b) As Fig. 12.3(a) except after unrestrained swelling under water; *p* 18.6 μm, θ 42/33°. The structure is more open and the 'particles' smaller. 984.8A/11A, Cambridge.

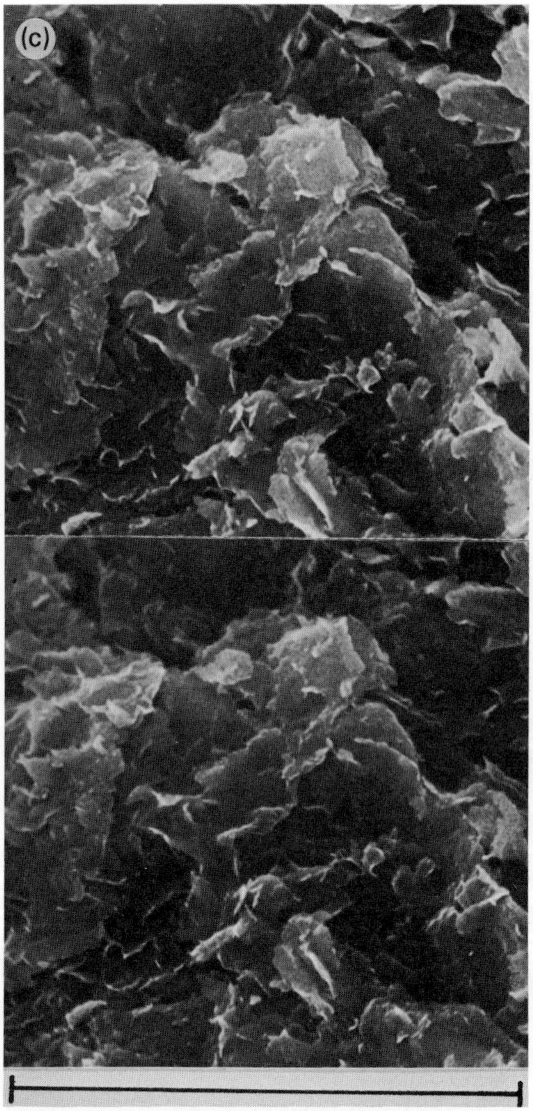

Fig. 12.3 (c). As Fig. 12.3(a) except after swelling under 100 kPa; *p* 15.2 μm, θ 51/42°. The structure is similar to Fig. 12.3(a) except that the voids are smaller and the particle edges more clearly defined; this indicates that the intra-domain voids have expanded at the expense of the inter-domain voids. 1116.22/23, Cambridge.

Fig. 12.4. Felspar grains (a) from core stone of granite boulder, (b) from weathered shell, from Middle Old Red Sandstone conglomerate from near Clunas, Nairnshire; hand-picked; ultrasonic cleaning in distilled water; shadowed carbon replicas; TEMs. Note the sharpness of the edges and corners in (a) and their rounding in (b). Supplied by W. J. McHardy. (From Wilson, Bain and McHardy 1971.) 70.453 and 70.465, W. J. McHardy, Macaulay Institute.

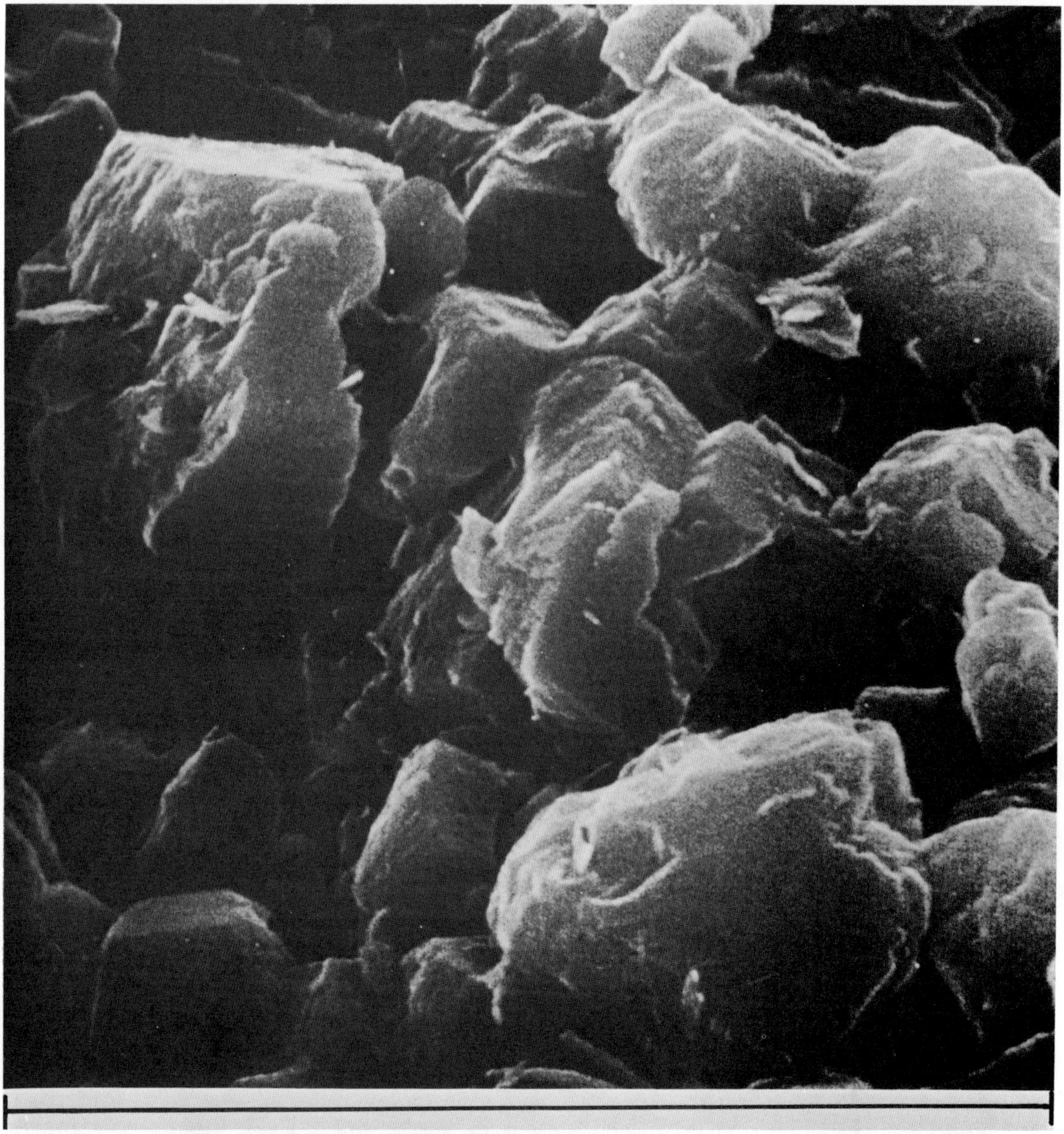

Fig. 12.5. Kaolinite stacks in Oxford Clay; undisturbed; air-dried; fracture surface; 20 nm Au; SEM, *p* 15 μm, θ 33°. This seems to show a mild case of irregular edge attack. (From Tovey, N. K. (1971). A selection of scanning electron micrographs of clays. Cambridge University Engineering Department Technical Report CUED/C—SOILS/TR5.) 395.37, Cambridge.

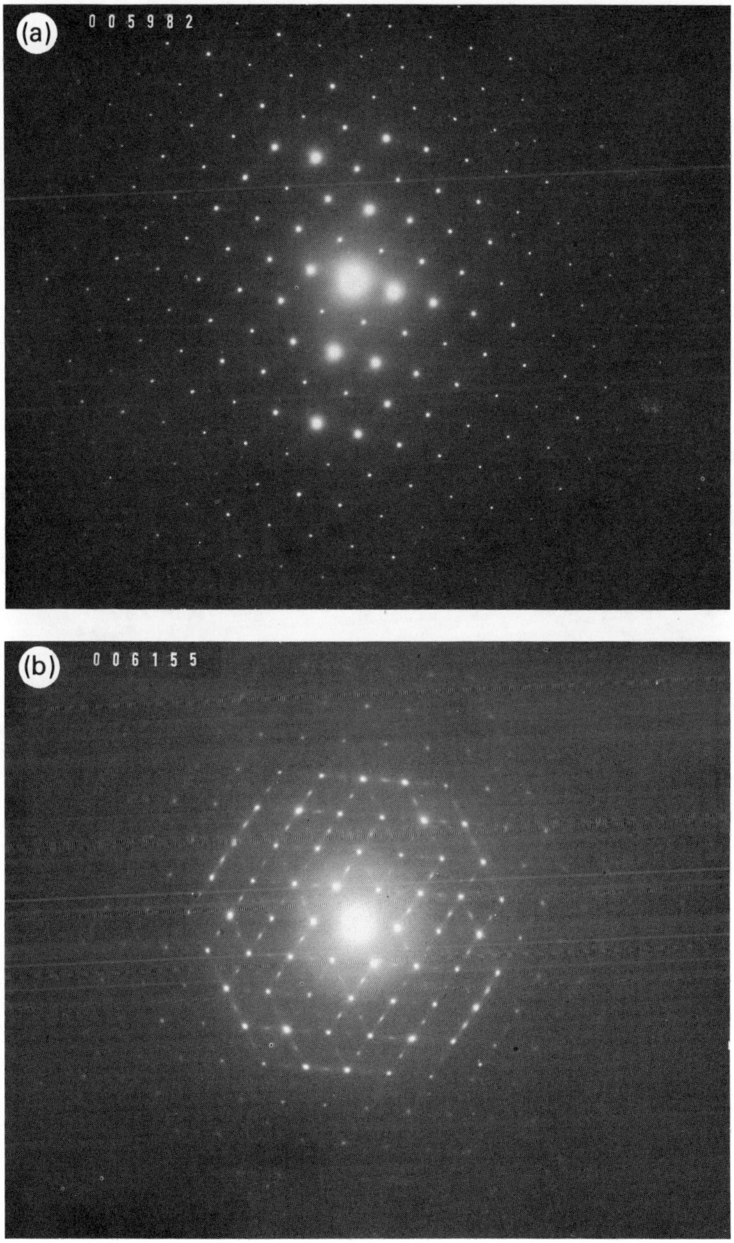

Fig. 12.6. Microcrystalline muscovite; potassium extracted from interlayer with 0.1 N $BaCl_2$ at 98 °C; converted to Ca-form with $CaCl_2$; dispersed particles; electron diffraction patterns: (a) natural; (b) Ca-form. In the Ca-form, the extra spots are superlattice reflections, almost certainly owing to the interlayer cations, possibly owing to a real superlattice, more probably owing to intracrystalline domains with primitive symmetry (as opposed to the face-centred symmetry of the K-form). There are probably three sets of domains, aligned in the three 'hexagonal' directions; the alignments may vary from layer to layer; lengthening of the extra spots suggests that the domains are 4–9 nm ϕ. Supplied by H. Kodama. (From Kodama 1977.) 5982 and 6155, Agriculture Canada.

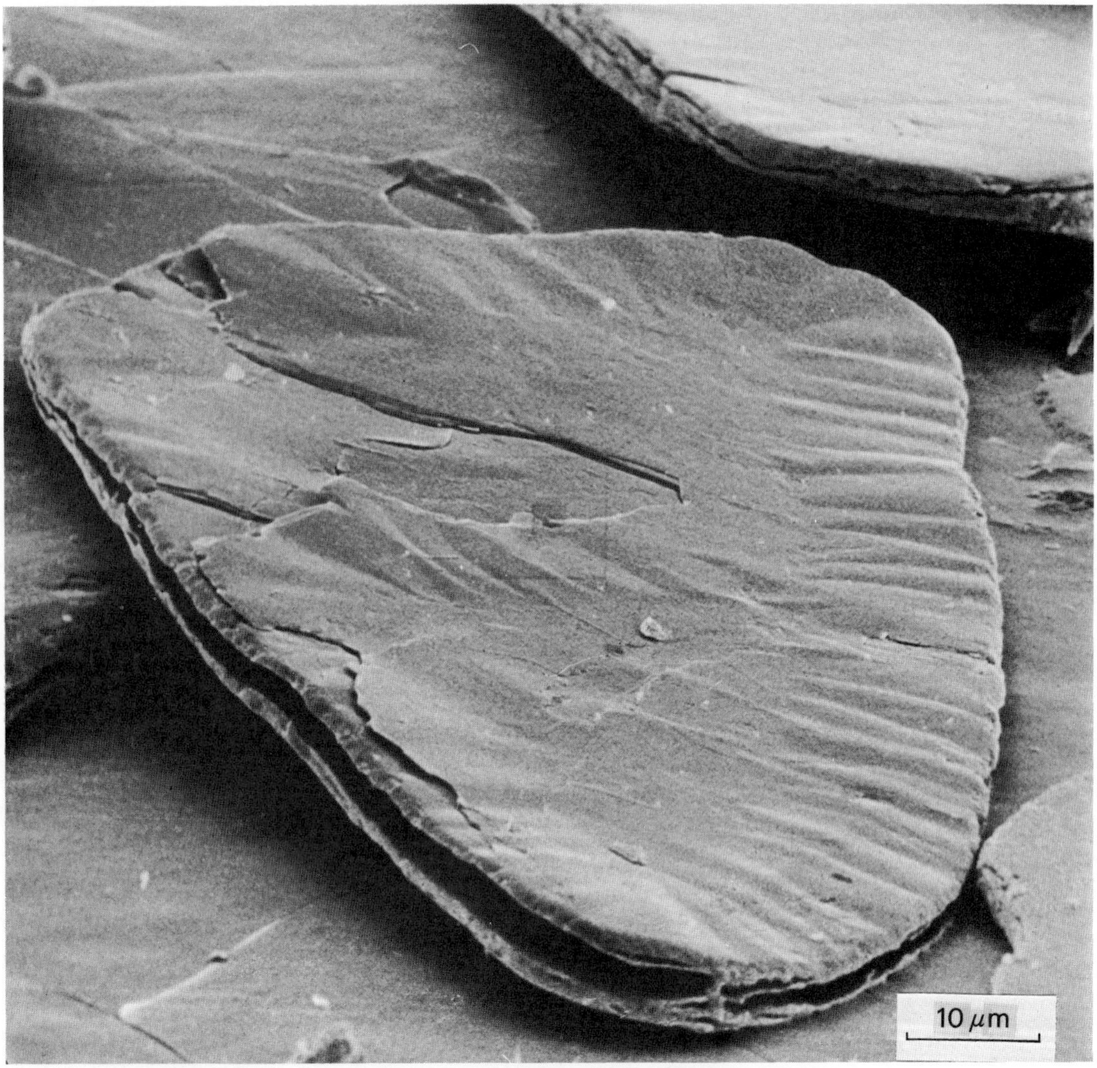

Fig. 12.7. Large biotite flake after partial removal of interlayer K by $BaCl_2$; SEM. Note the splits and wrinkles, which extend part-way into the particle from the edges. Supplied by H. Graf von Reichenbach. (From von Reichenbach 1972.) C. Samtleben, University of Kiel.

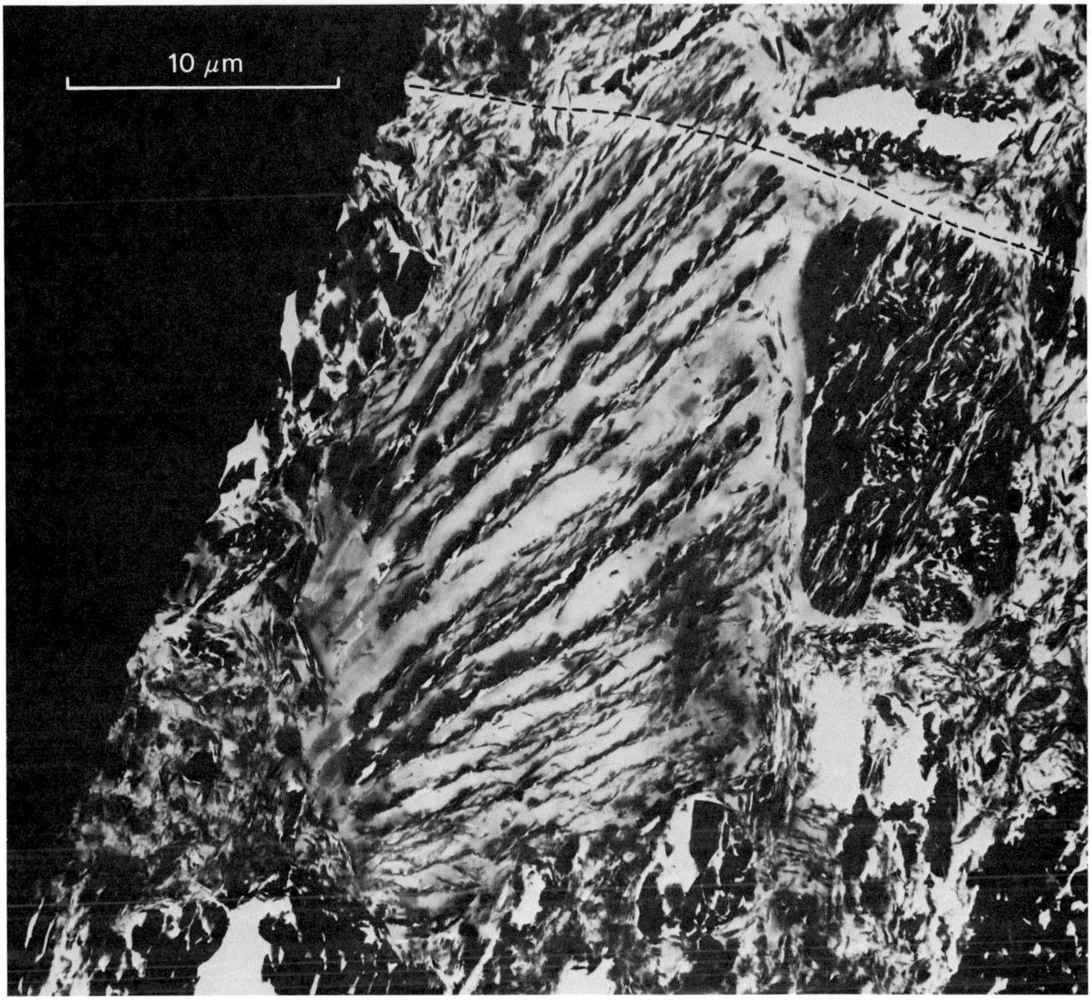

Fig. 12.8. Fen Clay; undisturbed sample; alcohol, acetone, Vestopal W; ultrathin section, vertical section; TEM. The layered feature, 33 × 16 μm, is thought to be a 'packet' formed from a silt particle which has been chemically weathered in situ; note (i) the Vestopal between the layers, indicating that the separation is not sectioning damage, (ii) the sub-micrometre particles within the packet, and (iii) the alignment of the plates around it. The darker feature to the right, 15 × 10 μm, is probably a silt particle partly shattered by microtomy; but note the irregular foliation. The white mark across the top of the field of view and near the scale mark is a knife mark. Same sample as Fig. 9.3. (From Stirling, J. M. (1973). Structure of Fen Buttery Clay. B.Sc. Report, University of Glasgow.) 3474, P.S., Glasgow, 338.5.1.

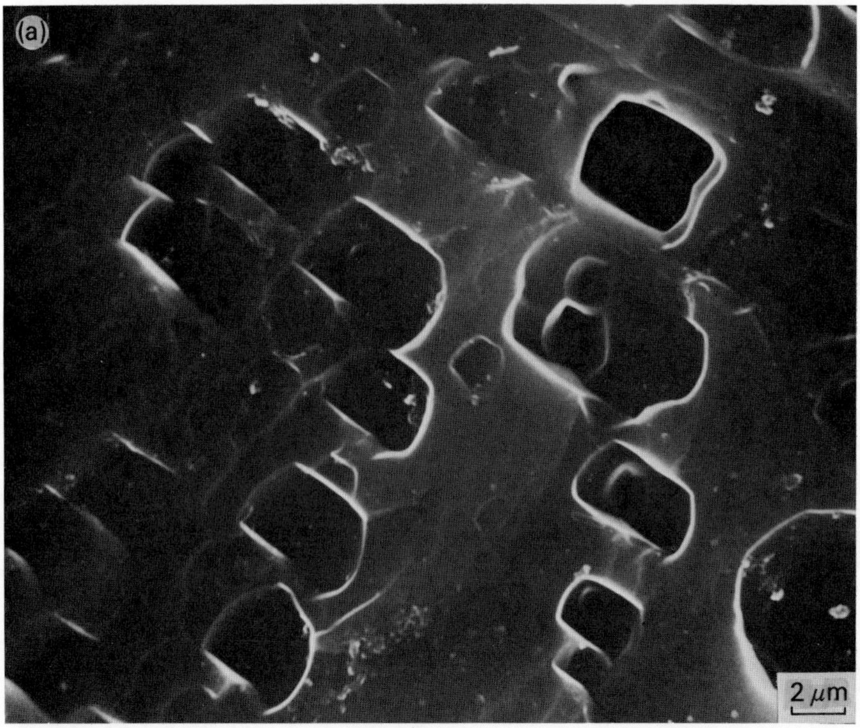

Fig. 12.9. SEMs showing formation of halloysite from plagioclase via an intermediate reaction product. (a) Parallel development of square-shaped etch pits. (b) Amorphous thin layer with irregular holes, Tazaki's stage or class Ht 2–1. (c) Spherical halloysite, up to 0.2 μm ϕ, forming from the amorphous layer, Ht 2–2. Supplied by K. Tazaki. (From Tazaki 1978*b*.)

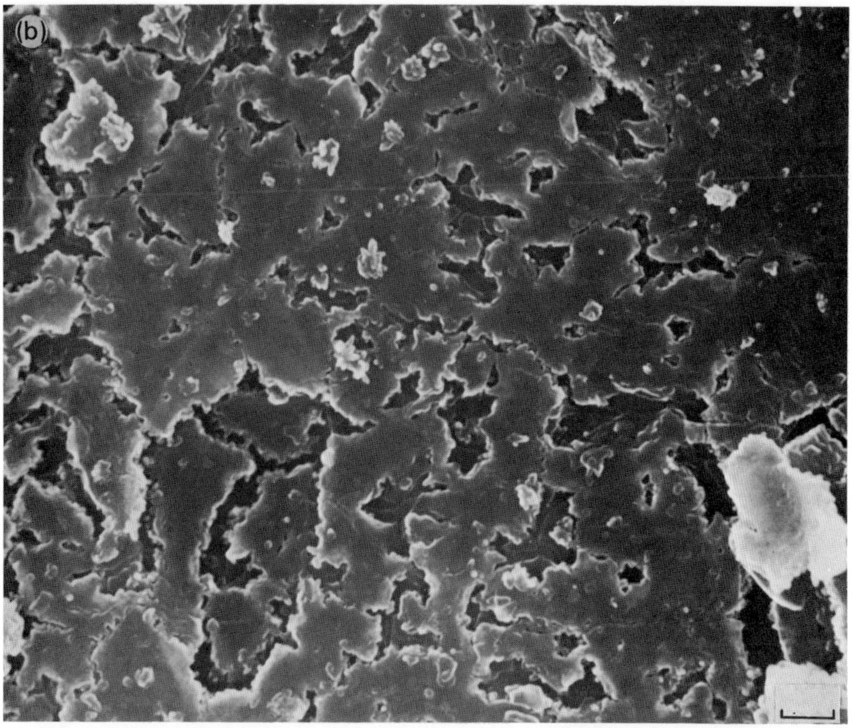
(b)

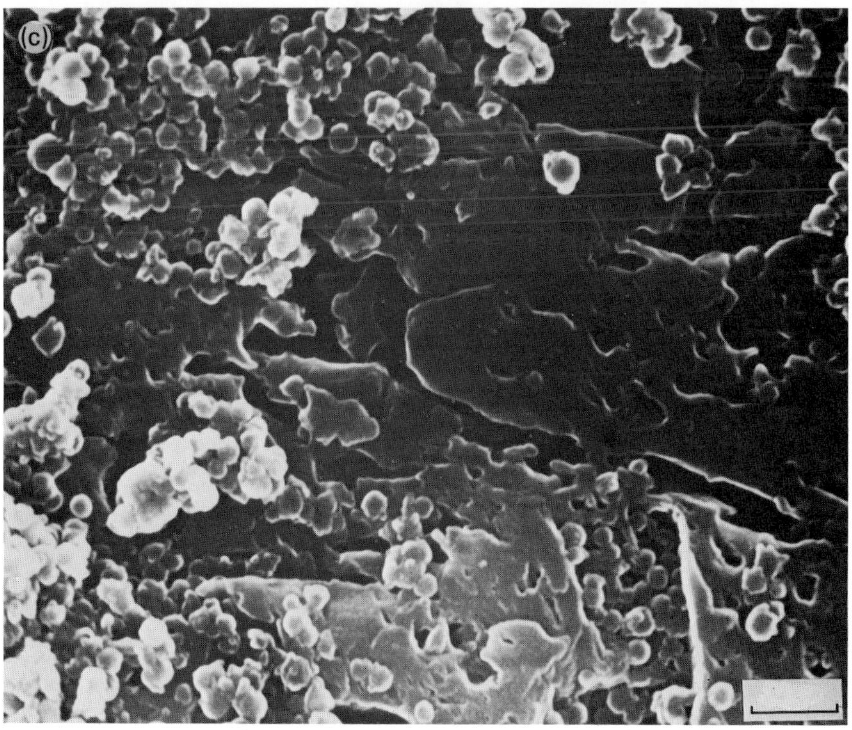
(c)

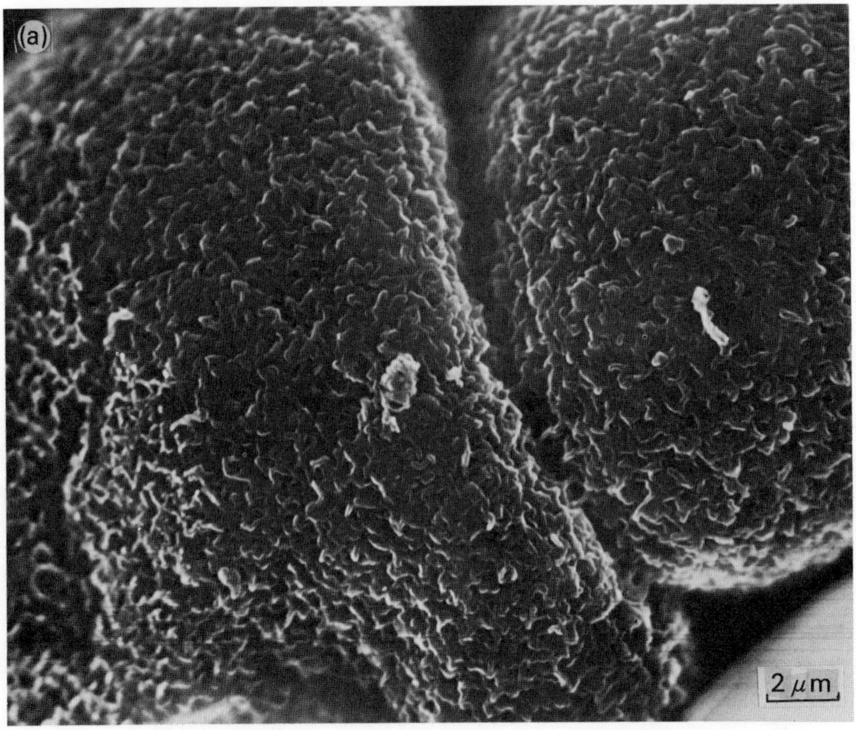

Fig. 12.10. SEMs showing further development of halloysite from plagioclase. (a) 'Walnut' shaped halloysite forming on a swollen surface, Ht 2–3. (b) The same, on a smooth surface, Ht 2–3. (c) Crinkly halloysite forming globules with a honeycomb-like surface, Ht 3–1, cf. Fig. 11.6. Supplied by K. Tazaki. (From Tazaki 1978*b*.)

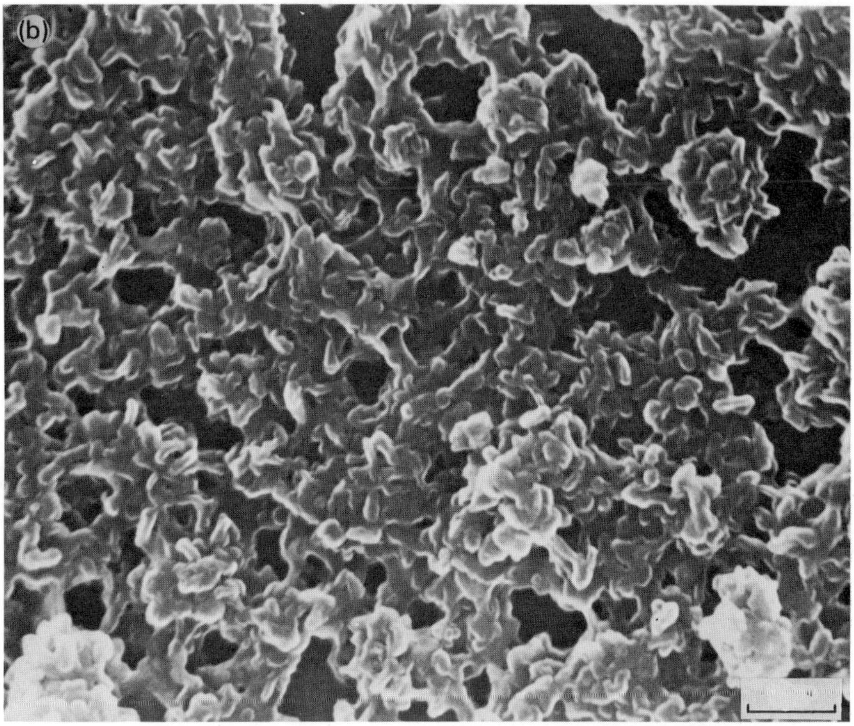
(b)

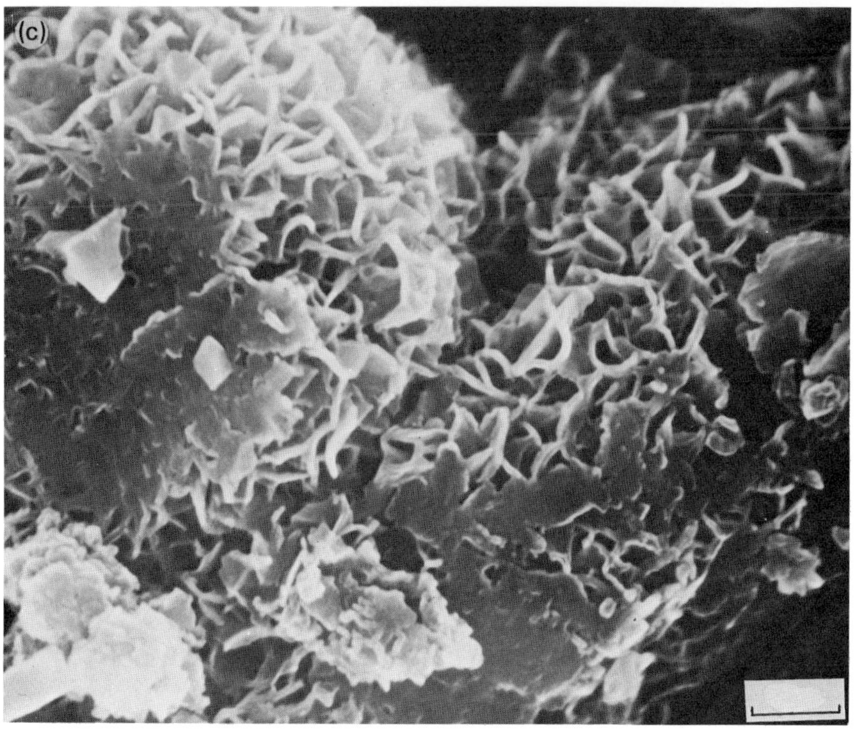
(c)

Fig. 12.11. SEMs showing formation of gibbsite from plagioclase via an intermediate reaction product. (a) The incipient stage: upper right, exsolution laminae in the plagioclase; centre, an amorphous thin layer; scattered, gibbsite crystals forming. (b) Gibbsite crystals forming from the thin layer. (c) Gibbsite crystals. Supplied by K. Tazaki. (From K. Tazaki 1976, 1978*a*.)

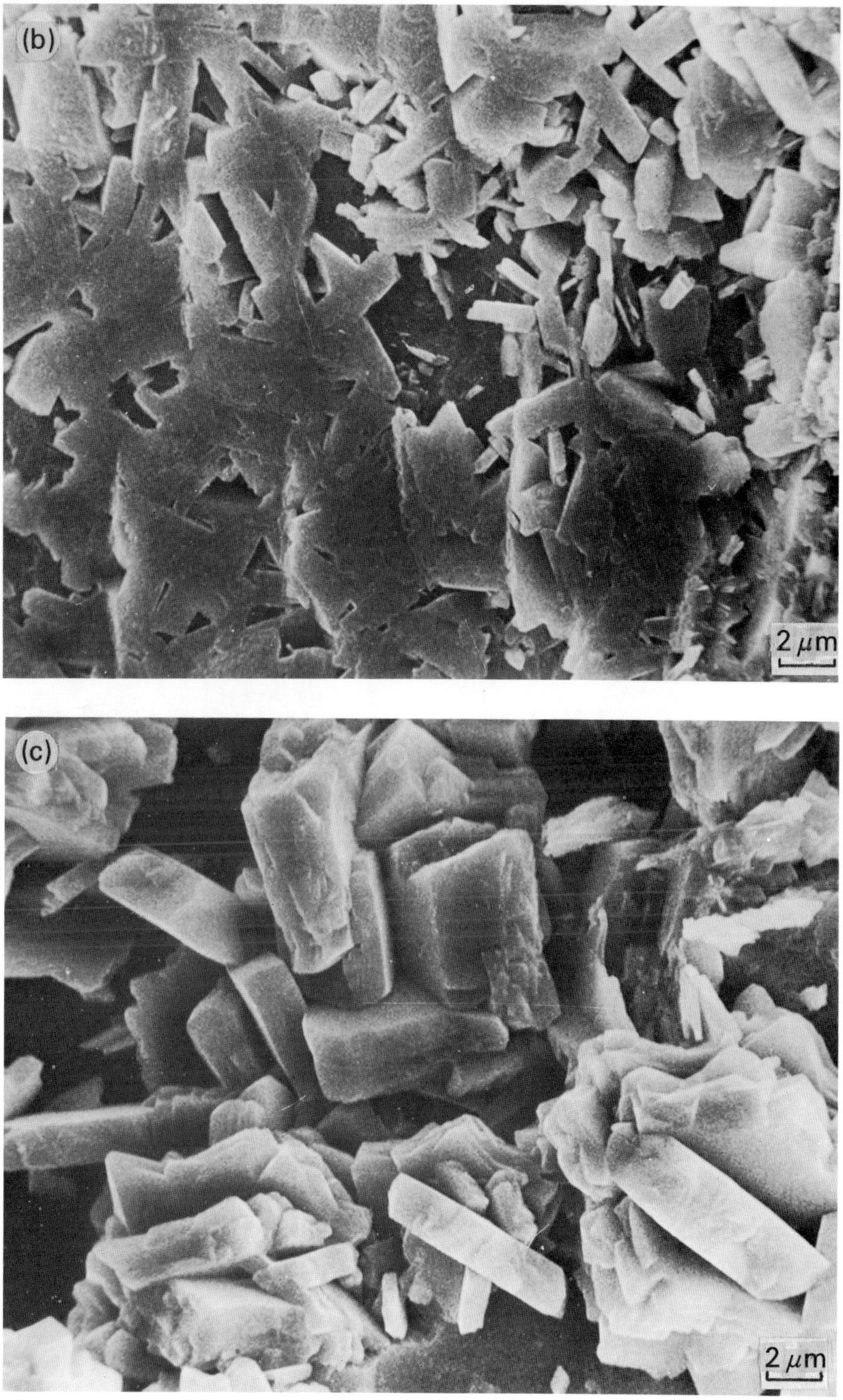
(b)
2 μm
(c)
2 μm

Fig. 12.12. Lateritic bauxite with relict macrostructure from Panchpatmali Plateau, Orissa, India; fracture surface; C, Au–Pd; SEM. Strong leaching high above the water table has left open cavities and a framework of large well-developed gibbsite with small irregular hematite and goethite. Supplied by G. Bárdossy. (From Bárdossy *et al.* 1978.) Hungarian Aluminium Corp.

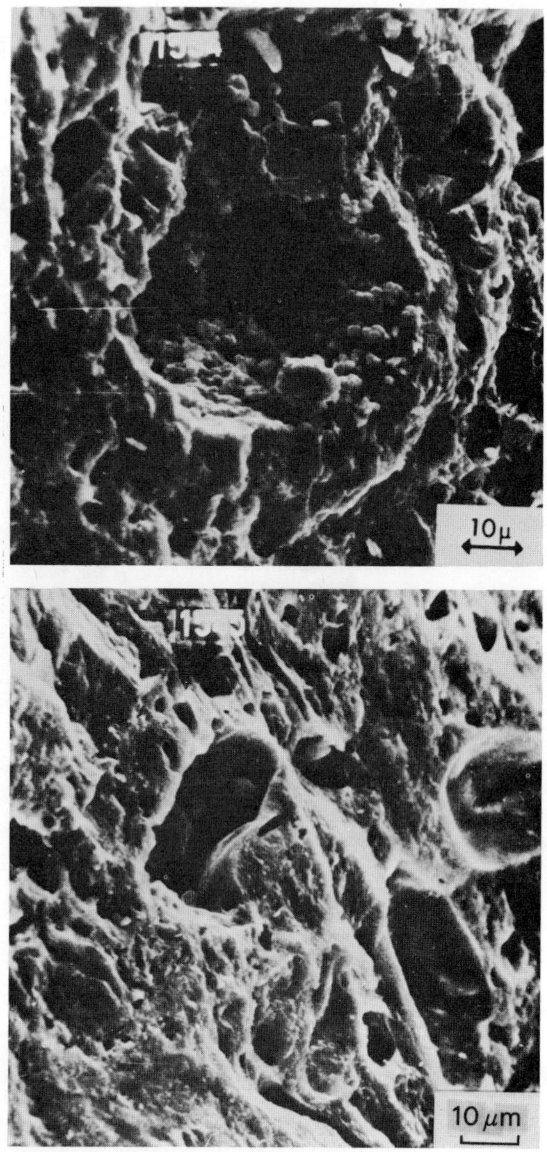

Fig. 12.13. Lateritic bauxite with pisolitic macrostructure from near Bamako, Mali; details as Fig. 12.12. Redistribution of the weathering products after leaching has left a collomorphous-spongy space-filling with curvilinear voids.

13 Erosion

The micrographs shown illustrate the only electron microscopic study of erosion known to us.

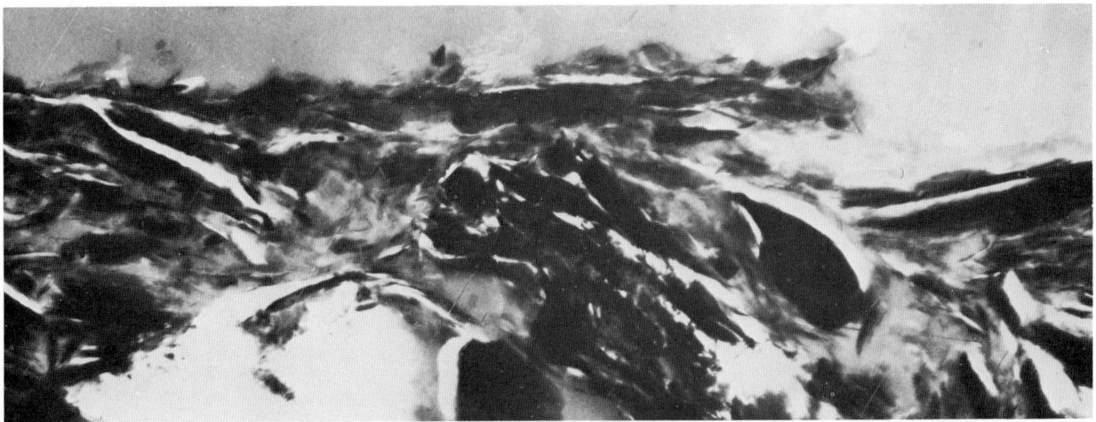

CV

← H →

CV

← H →

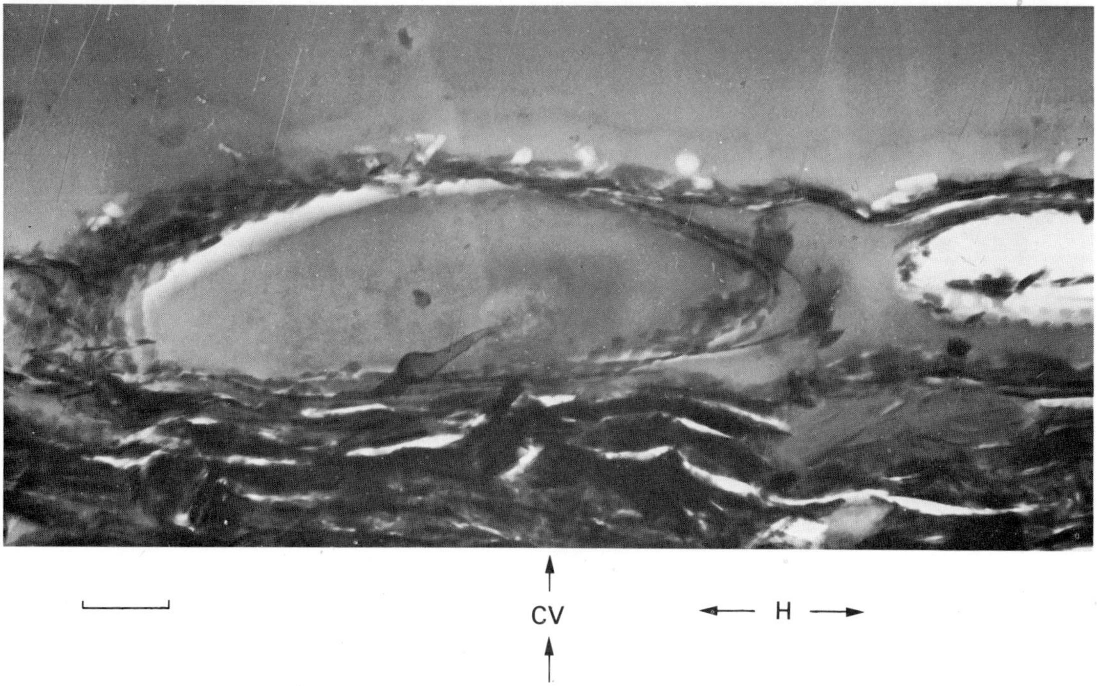

Fig. 13.3. Same sample and preparation as Fig. 13.1 from the middle beach. The oval feature is (the remains of) a diatom, with another on the right. These look as if they are burrowing along just under the surface. 6262, P. A. Maclean, Glasgow, 348.3.2.1.V.

◁ **Figs. 13.1 & 13.2.** Fen Clay from Brancaster Beach, near Hunstanton; undisturbed samples, Kubiena boxes; methanol, acetone, Vestopal W; ultrathin sections, vertical sections, vertically cut; TEMs. These show cross-sections of the natural surface of the clay, which is being eroded. The lower micrograph is from the lower beach, near the low water level; note the random arrangement of the particles at the surface, the subsurface voids, and a tendency for small groups of particles to be eroded *en bloc*. The upper micrograph is from the middle beach, between the high and low water levels; note the parallel arrangement of the particles at the surface, which is thought to result from a different mechanism of erosion. 6237 and 7865, P. A. Maclean, Glasgow, 348.3.2.1.V and 348.6.2.6.V.

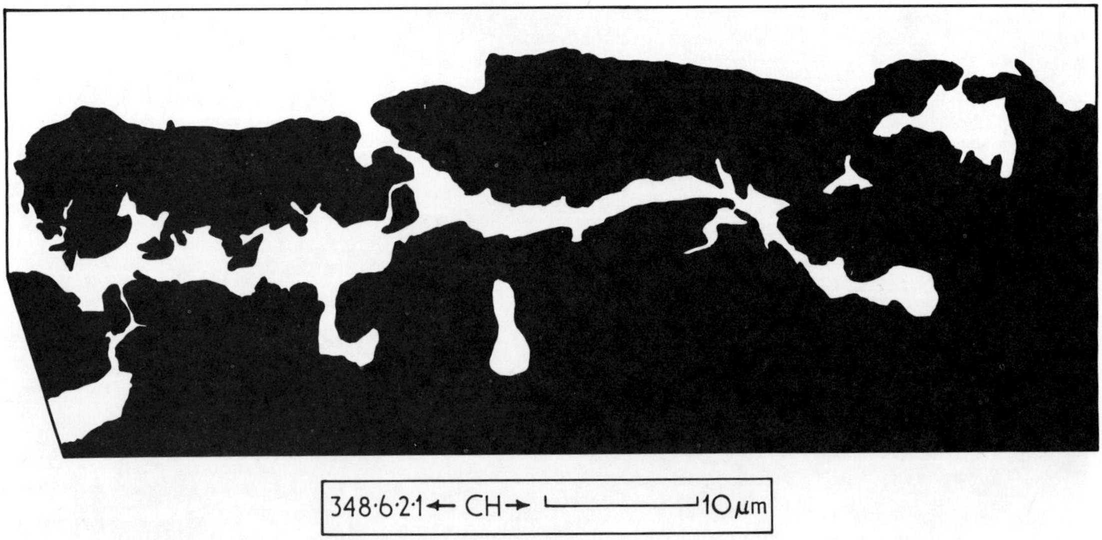

Fig. 13.4. Same sample and preparation as Fig. 13.2 from the lower beach; cross-section drawn from a series of micrographs of vertical ultrathin sections, horizontally cut. This illustrates the tendency for small groups of particles to be eroded *en bloc*. P. A. Maclean and P. Smart, Glasgow, 348.6.2.1.H.

14 Sand grains

Geologists have paid considerable attention to correlating the surface features seen on sand grains with their geological history. The first part of this Section summarizes this work. The last Figure illustrates an engineering study of the liquefaction of sand.

Both the overall shape and the microtextural features of a sand grain can give a general indication of its environment and age. Recently formed glacial grains tend to be elongate and/or irregular in shape with a relatively smooth microtexture. With time, mechanical and chemical action round the grains and endow them with characteristic microtextural features. For quartz grains, the most important processes are:

1. Mechanical action.
   1.1. Aeolian action.
   1.2. Sub-aqueous action.
   1.3. Pressure solution.
   1.4. Melting during collisions.
   1.5. Glaciation.
2. Chemical action.
   2.1. Etching.
   2.2. Solution-precipitation.
   2.3. Crystal growth.

A more detailed discussion of these textures is given by Krinsley and Doornkamp (1973); see also Section 12.

The susceptibility of a sand to liquefaction depends on factors such as mean particle size, particle size distribution, particle shape, particle surface microtexture, electrochemical effects, and the voids ratio. This susceptibility increases with increasing voids ratio and elongation of the particles and decreasing particle size. Under some conditions, the microtextural features can affect the resistance to stress, especially when the effective stress is lowered by a high pore water pressure.

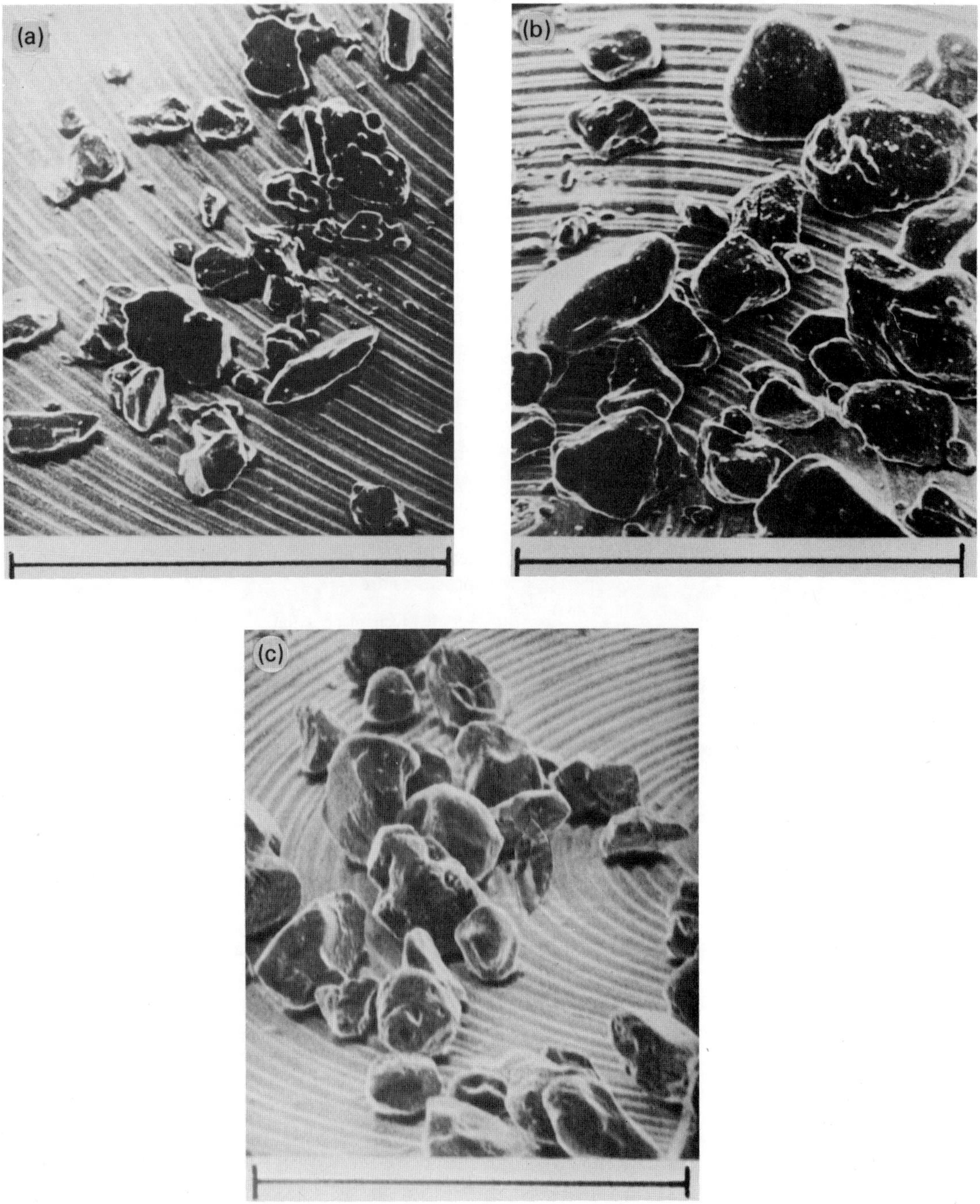
(a)
(b)
(c)

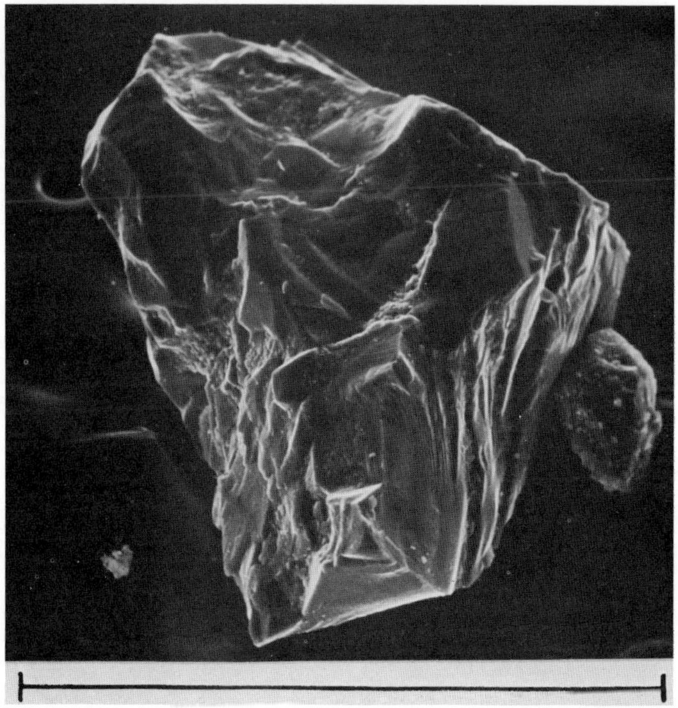

Fig. 14.2. Sand grain recently deposited from a glacier into a Norwegian fjord; 20 nm Au; SEM, *p* 250 μm. The angular nature of this grain is indicative of glaciation. Taken and supplied by D. H. Krinsley, Arizona State University.

◁ **Fig. 14.1.** Sand grains; scattered through a methanol–water mixture onto specimen stubs thinly smeared with Araldite; 25 nm Au; SEMs, *p* 850, 735, 735 μm respectively, θ 60, 51, 60° respectively. (a) Recently deposited from a glacier, Valgrinda, Norway; note the irregular shapes. (b) From the Rhine Delta, Holland; the grains are more rounded. (c) From Permian deposit, Ottawa; many grains are well rounded. Samples supplied by Binnie & Partners. 647.10A, 673.21, and 658.5, Cambridge.

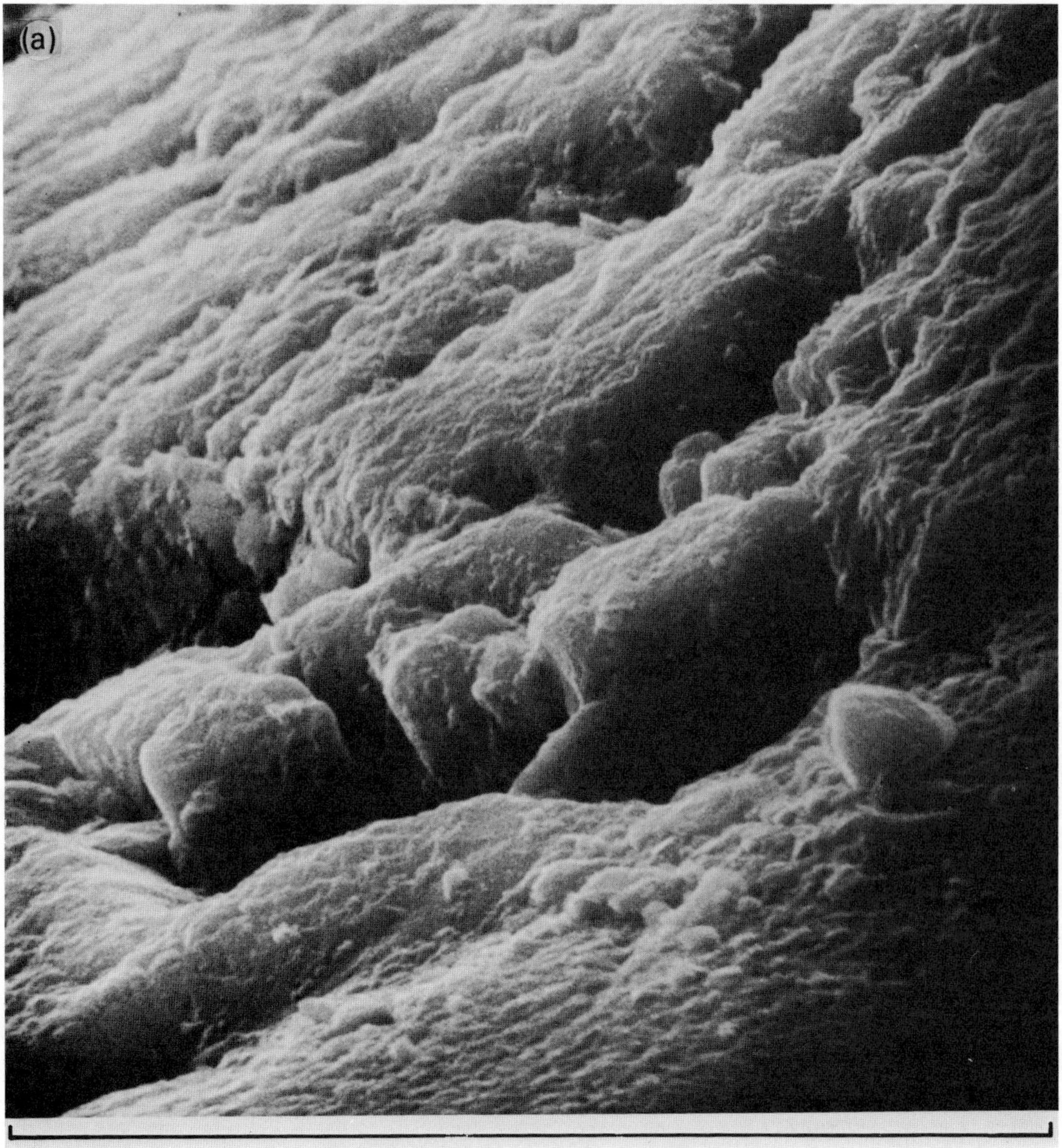

Fig. 14.3. Leighton Buzzard Sand, 14–25 BS sieve fraction; grains mounted individually; 20 nm Au; SEMs, *p* 18, 68 μm, *θ* 45°. (a) The series of parallel ridges or plates is characteristic of aeolian action in a non-desert environment; the depressions between the plates form along cleavage directions as a result of collisions. (b) The irregular patterns, which are only approximately V-shaped, and which are not oriented along crystallographic directions, are evidence of a moderate-energy sub-aqueous environment. 460.33 and 468.8A, Cambridge.

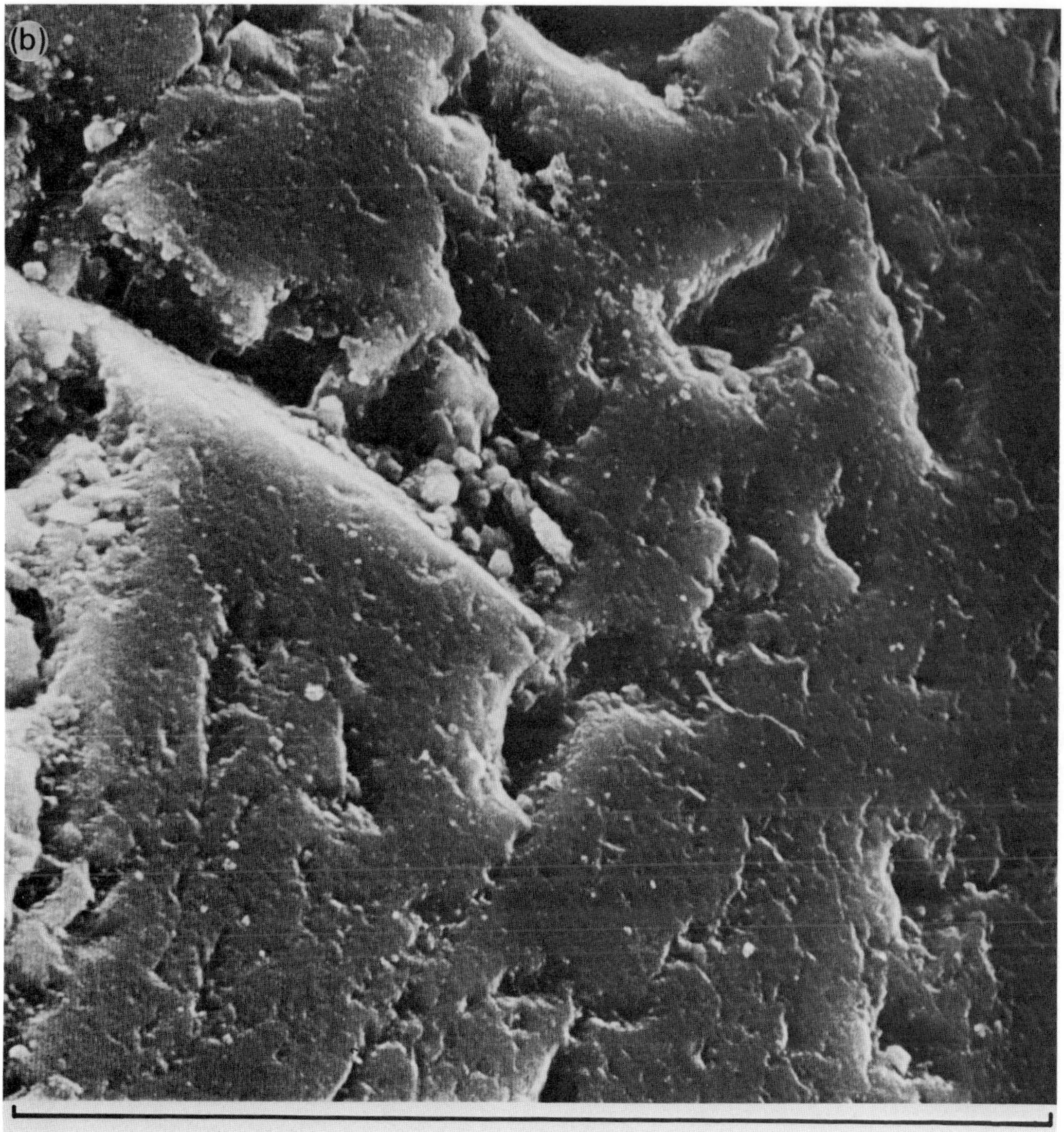
(b)

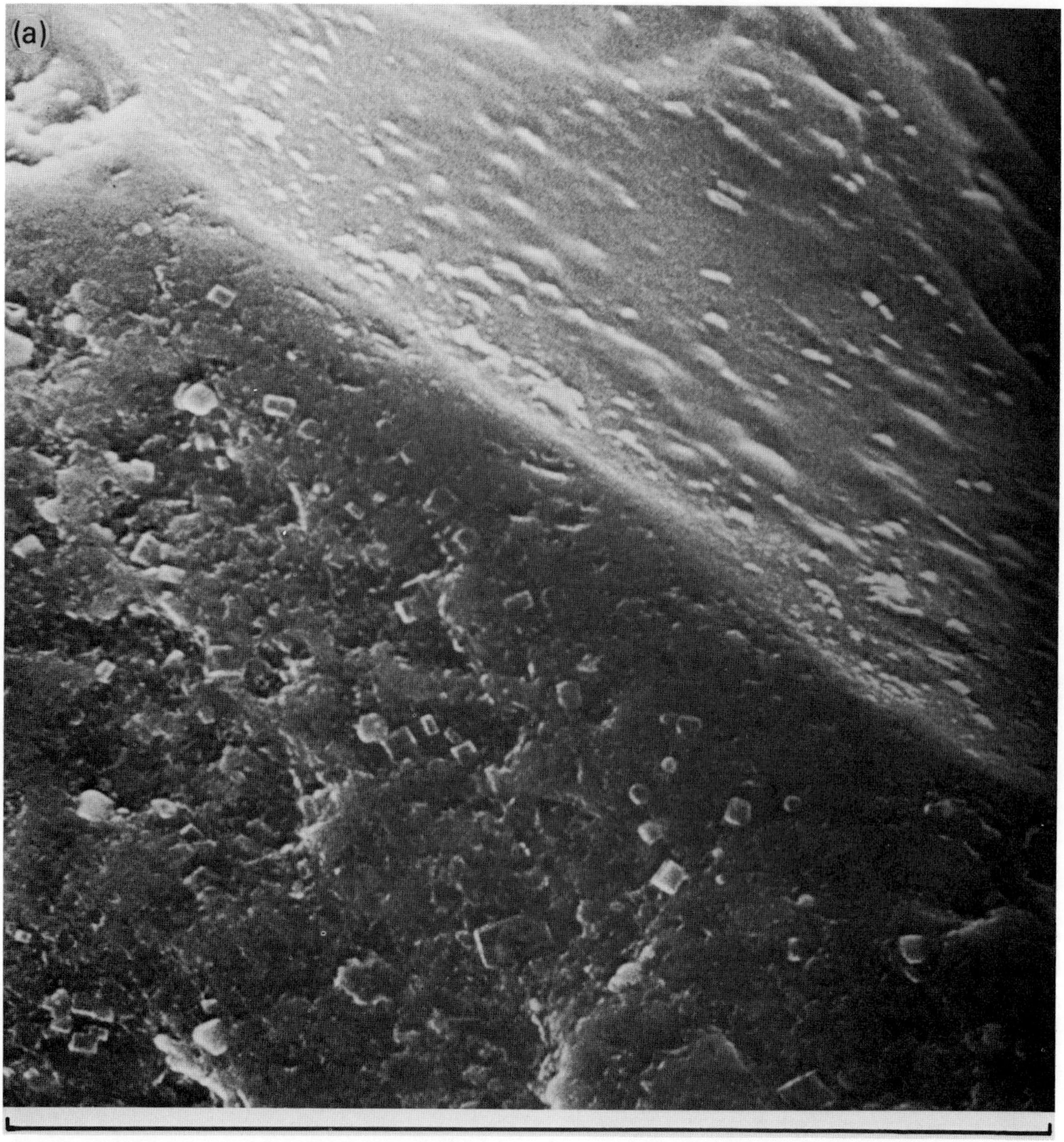

Fig. 14.4. Wash Sand from near surface at TF558237; 25 nm Au; SEMs, *p* 100, 107 μm respectively, θ 42°. (a) Early development of aeolian plates. (b) Extensive low-energy sub-aqueous mechanical erosion. Sample supplied by Binnie & Partners. 1068.9 and 1071.28, E. G. King, Cambridge.

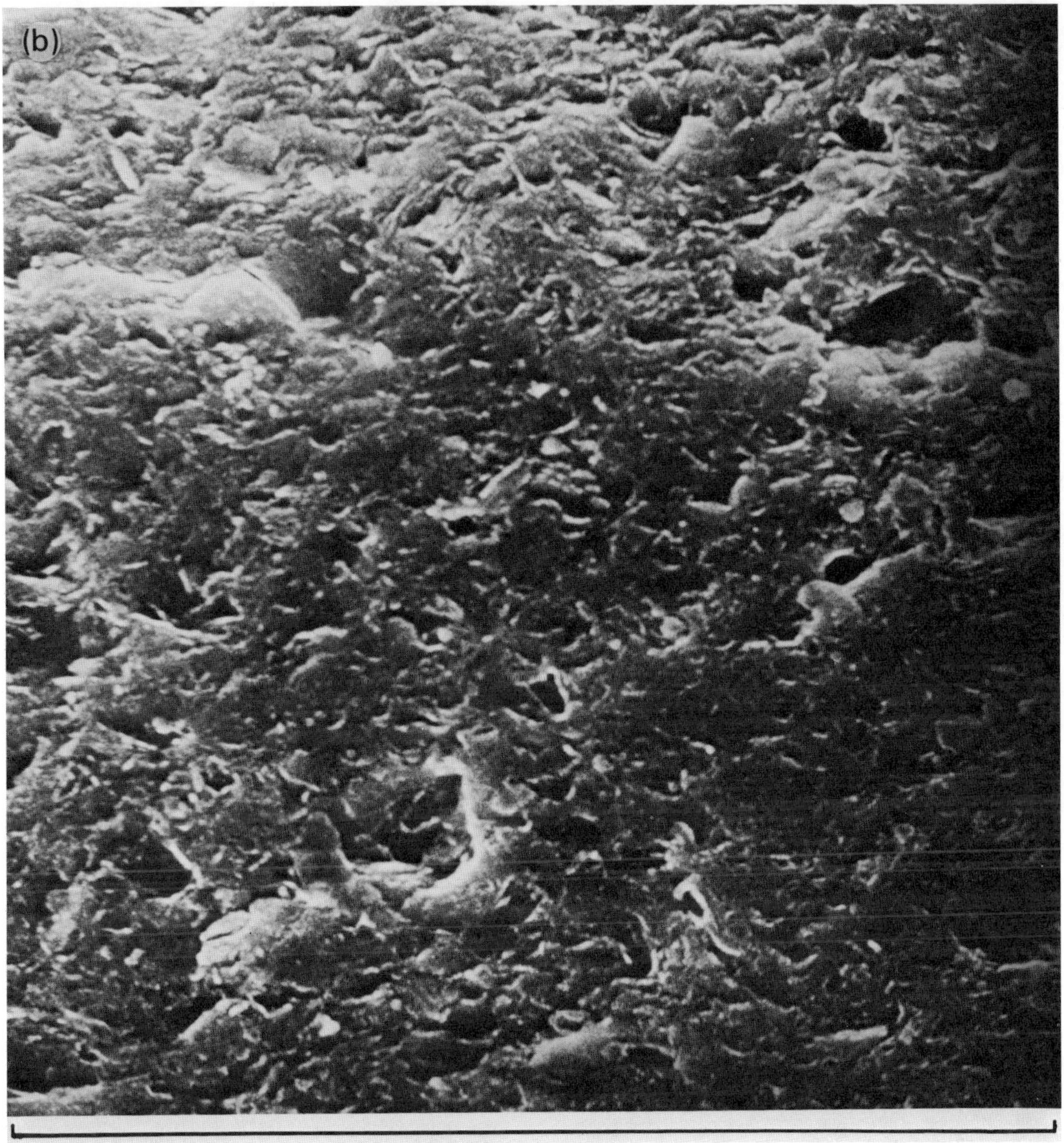
(b)

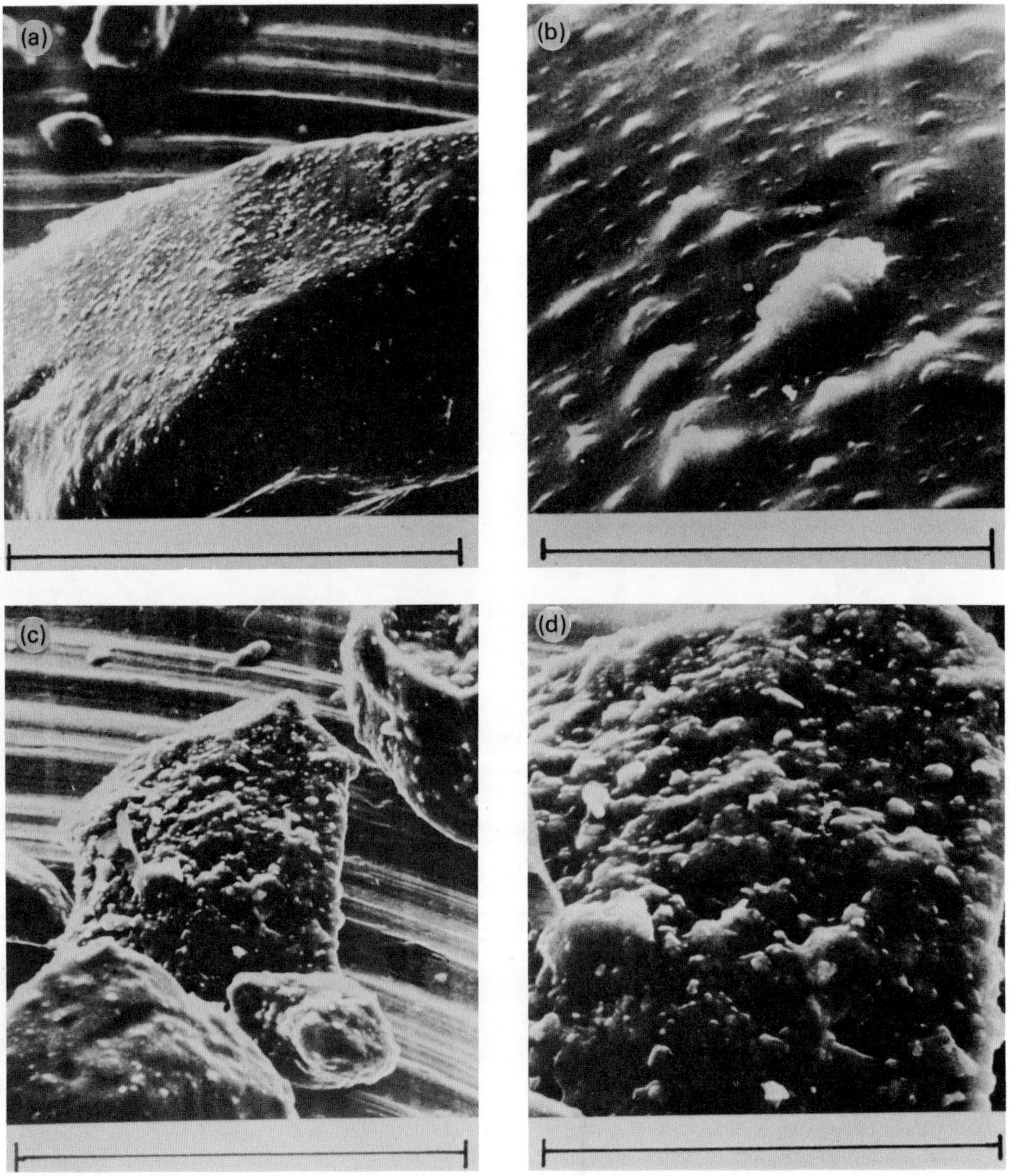
(a)
(b)
(c)
(d)

◁ **Fig. 14.5.** Sand from Rhine Delta, Holland; scattered through methanol water onto stubs smeared with Araldite; 25 nm Au; SEMs, *p* 200, 20, 200, 100 μm respectively, θ 42, 42, 51, 51° respectively. (a) and (b) Concentric micrographs showing the early development of aeolian type plates: the rounded nature of the plates indicates deposition of silica; note the similarity with Fig. 14.4 (a). (c) and (d) Almost concentric micrographs typical of rapid deposition of silica: small crystals of quartz growth are engulfed in amorphous silica. 673.19, 673.17, 674.22, and 674.23, Cambridge.

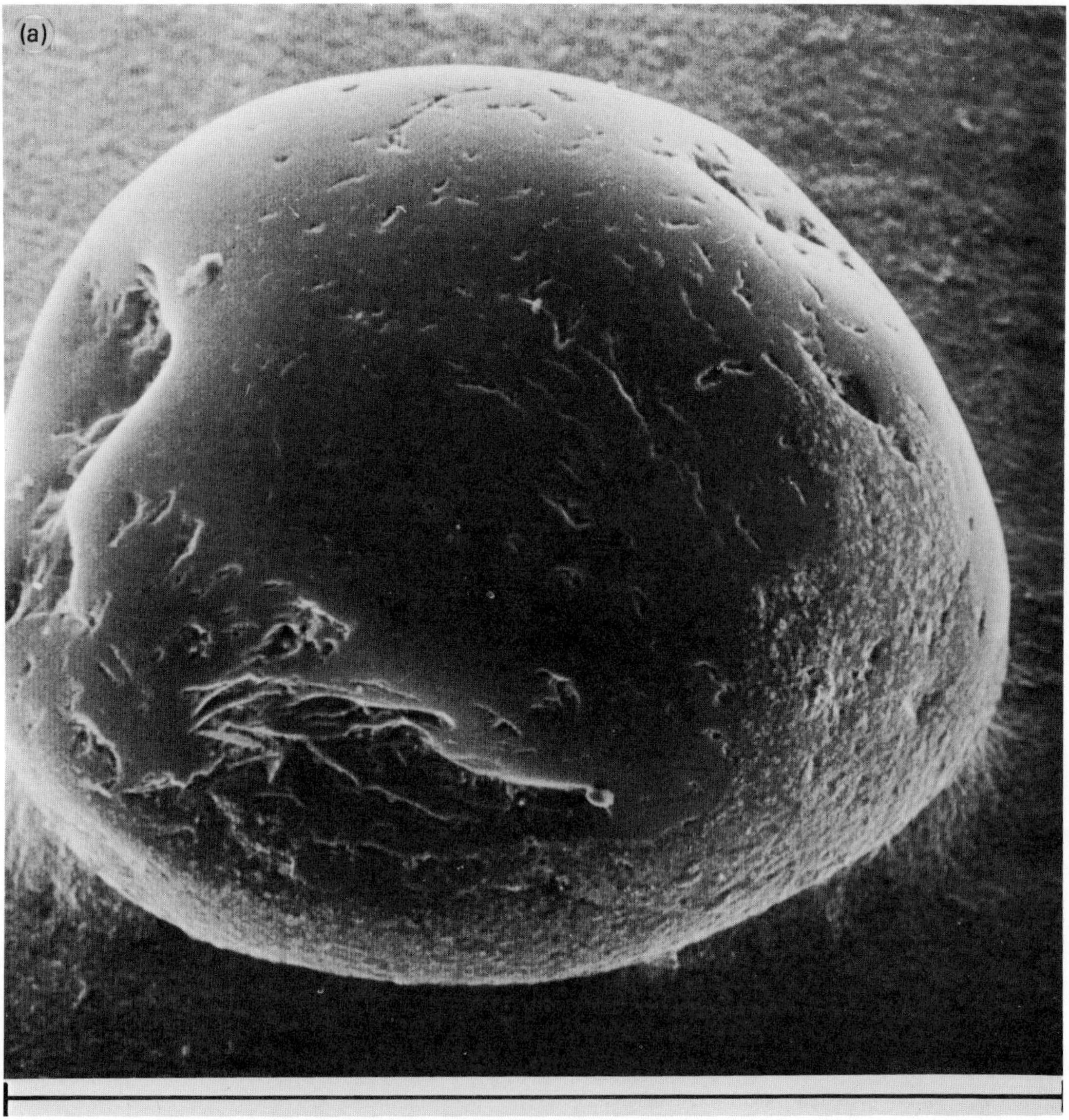

Fig. 14.6. Coastal dune sand from Iona; 25 nm Au; SEMs, *p* 168, 56 μm respectively, *θ* 50°. (a) shows rounding owing to mechanical action. (b) shows V-shaped patterns oriented along crystallographic directions, which indicate chemical etching. Sample supplied by D. H. Krinsley. 452.29A and 452.30A, Cambridge.

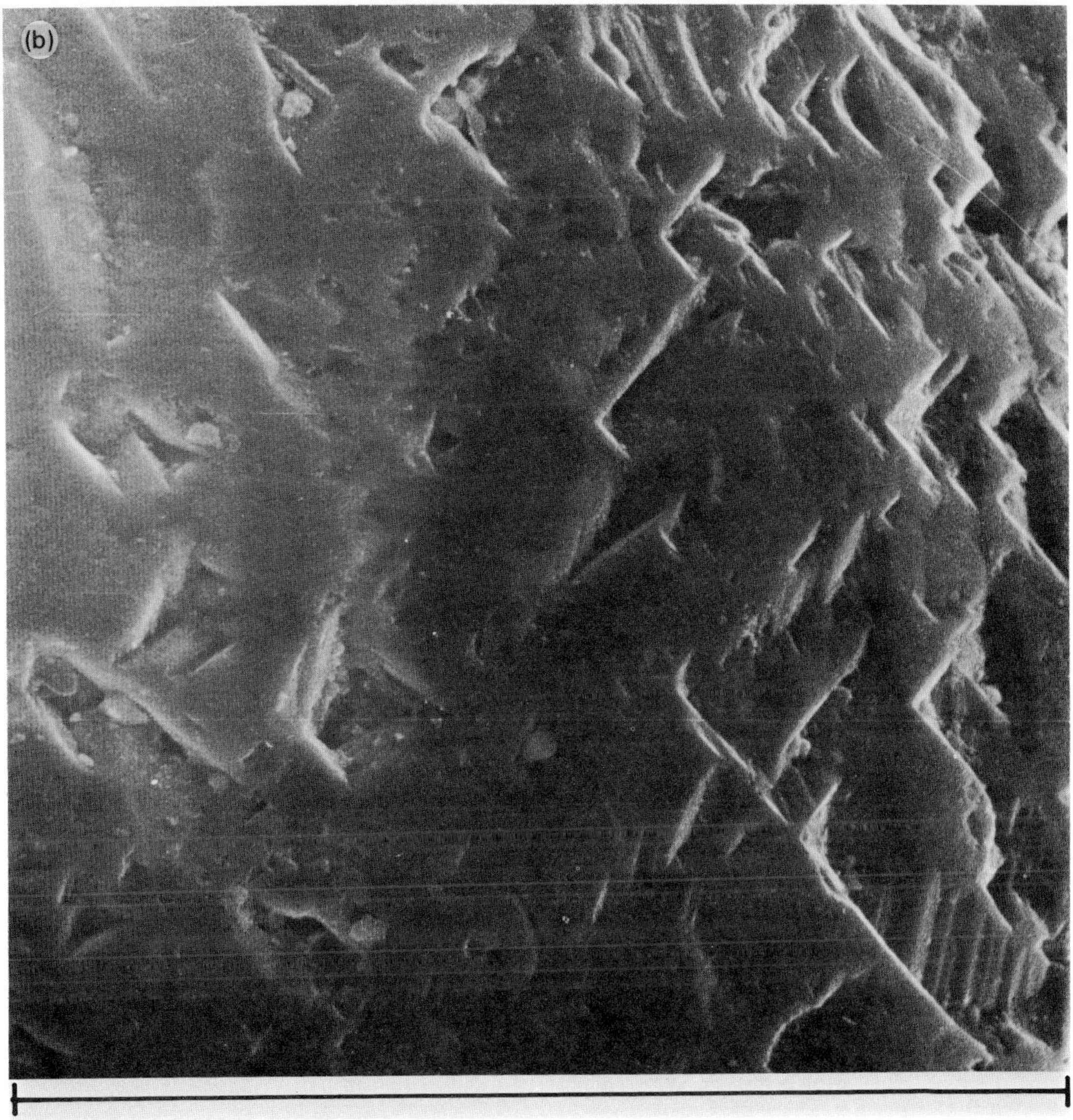
(b)

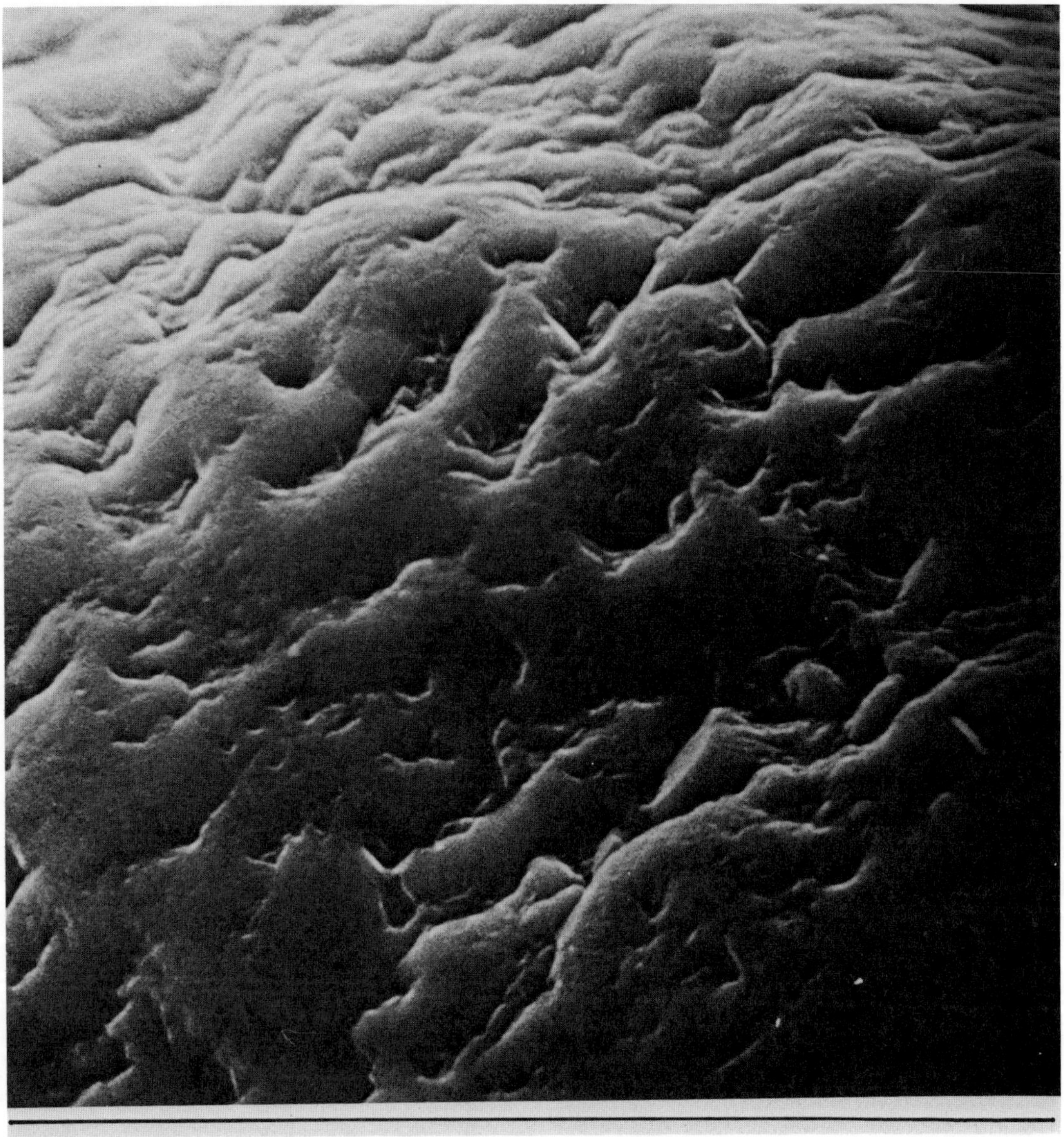

Fig. 14.7. Wash Sand from near surface at TF558237; 25 nm Au; SEM, *p* 76 μm, θ 42°. Etched surface, similar to Fig. 14.6(b), but modified by solution and reprecipitation of quartz. Sample supplied by Binnie & Partners. 1058.31, Cambridge.

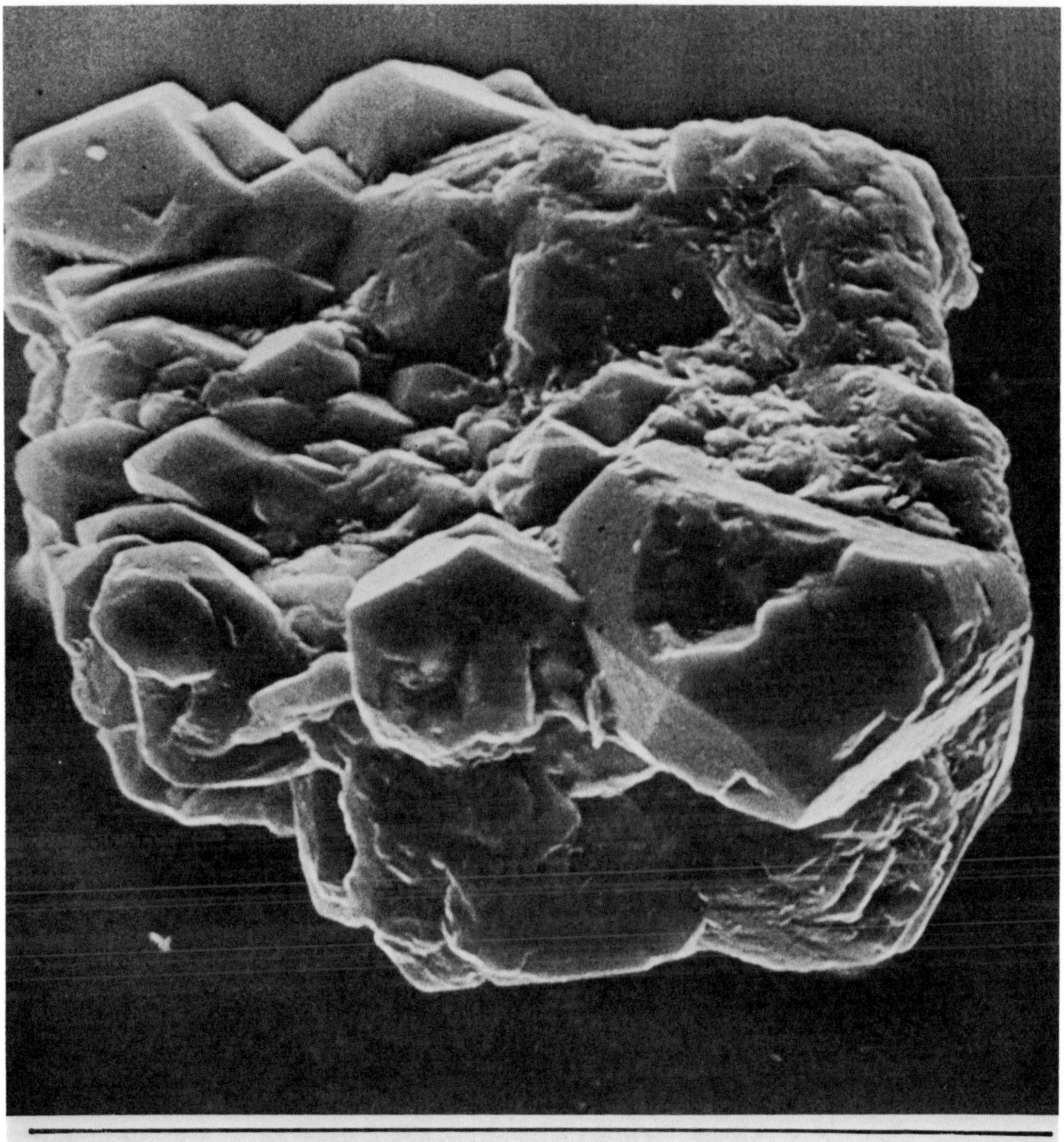

Fig. 14.8. Wash Sand from 8 m depth at TF558237; 25 nm Au; SEM, *p* 126 μm, *θ* 42°. This shows extensive crystal growth, which was common at this depth. Sample supplied by Binnie & Partners. 1050.15A, E. G. King, Cambridge.

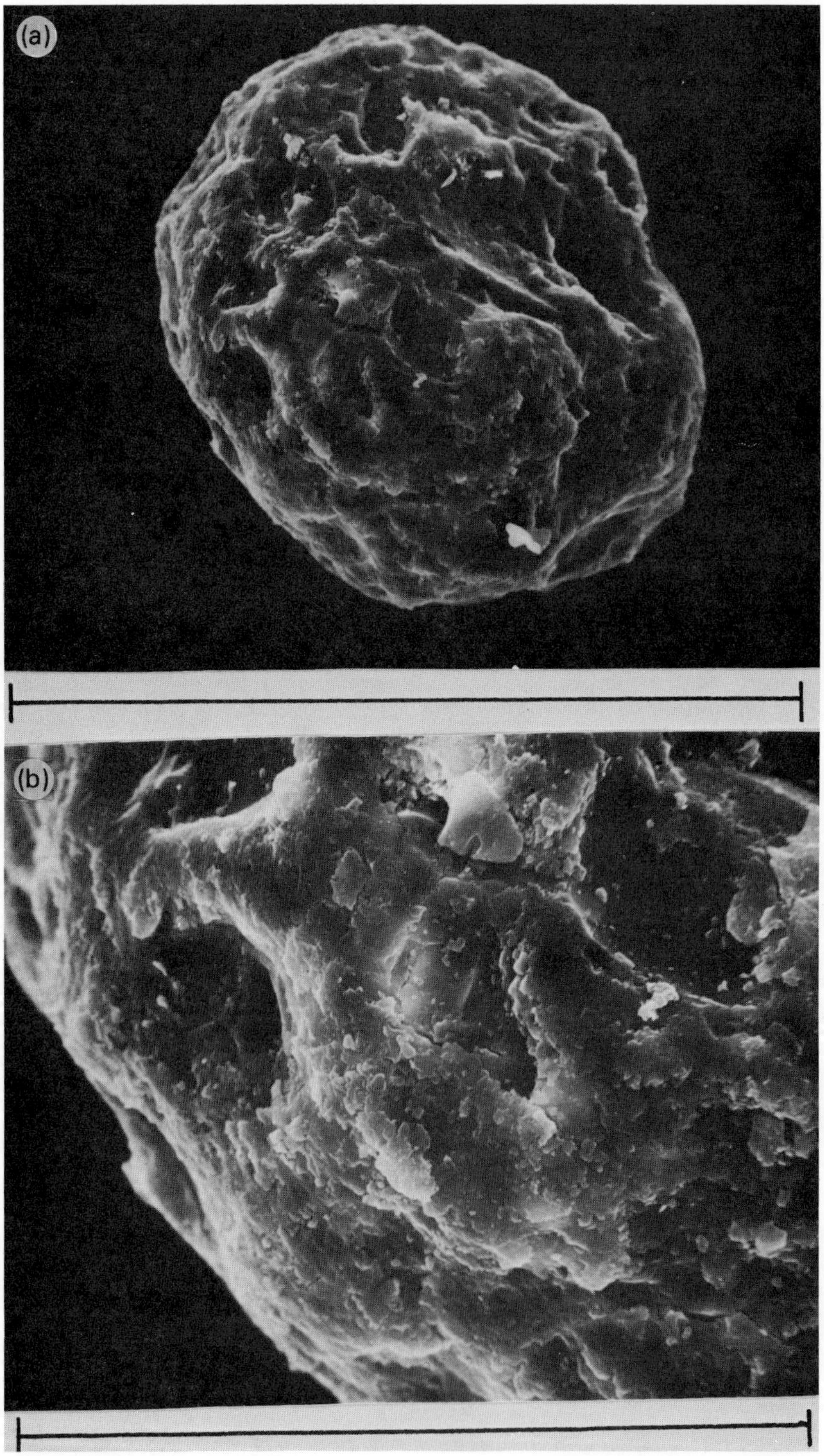
(a)
(b)

Fig. 14.10. Quartz grain tested as in Fig. 14.9; except at atmospheric pressure. The figure shows a montage of 1100 SEMs, and the original is 3.5 m ϕ. These high-magnification micrographs enabled the detailed microtexture over the whole grain to be studied; several different features were seen, and these were related to the crystallographic axes of the grain. (From Wellendorf, W. G. (1979). Aeolian microtextures on quartz sand. M. Sc. thesis, Arizona State University.) Arizona State University.

◁ **Fig. 14.9.** Grain of basalt after 20 minutes in an apparatus which simulated a Martian environment by blowing grains around at 32 ms^{-1} under reduced atmospheric pressure; 25 nm Au; SEMs, *p* 306, 121 μm respectively. The grain has become more rounded during this treatment, and the small plates tangential to the surface are caused by melting of parts of the grain by the dissipation of kinetic energy during collisions. Taken and supplied by D. H. Krinsley, Arizona State University.

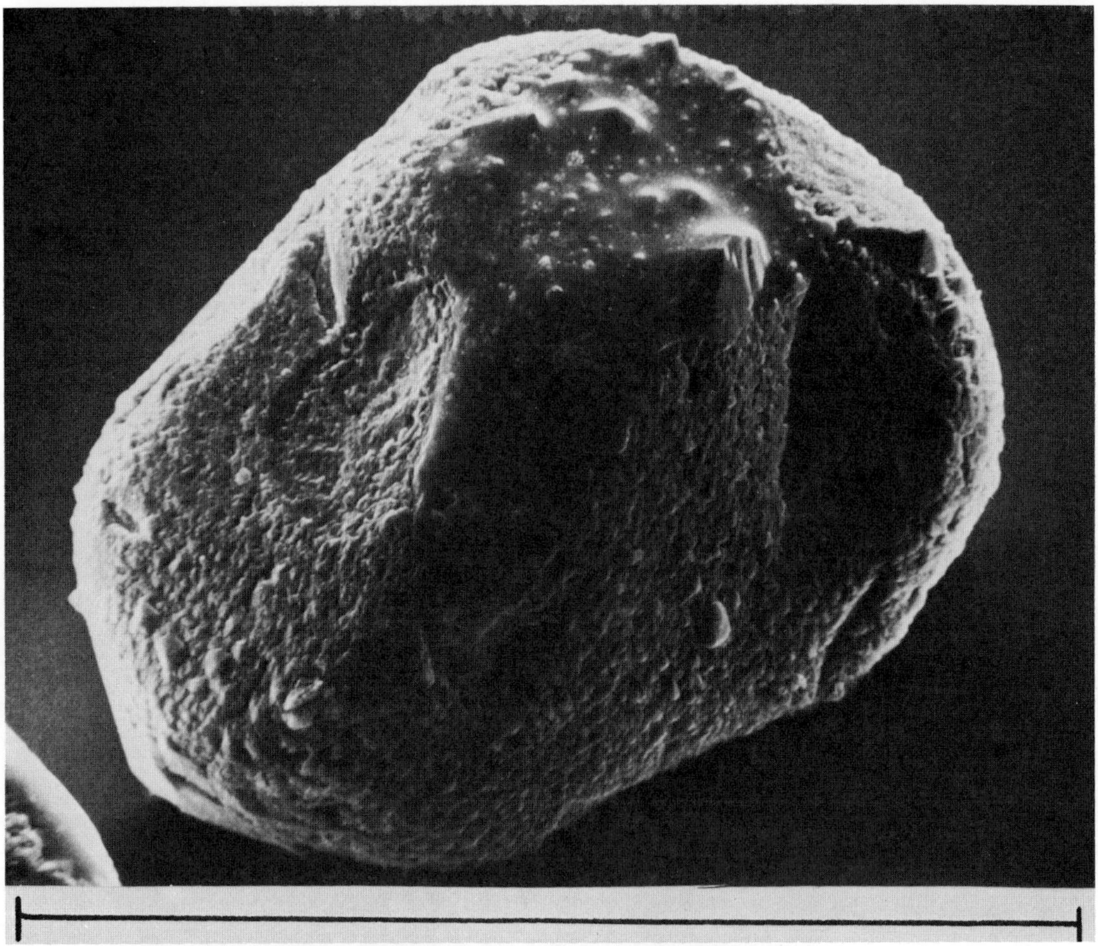

Fig. 14.11. Sand grain from Navajo Sandstone, 25 nm Au; SEM, *p* 520 μm. The dark smooth areas are evidence of pressure solution. Taken and supplied by D. H. Krinsley, Arizona State University.

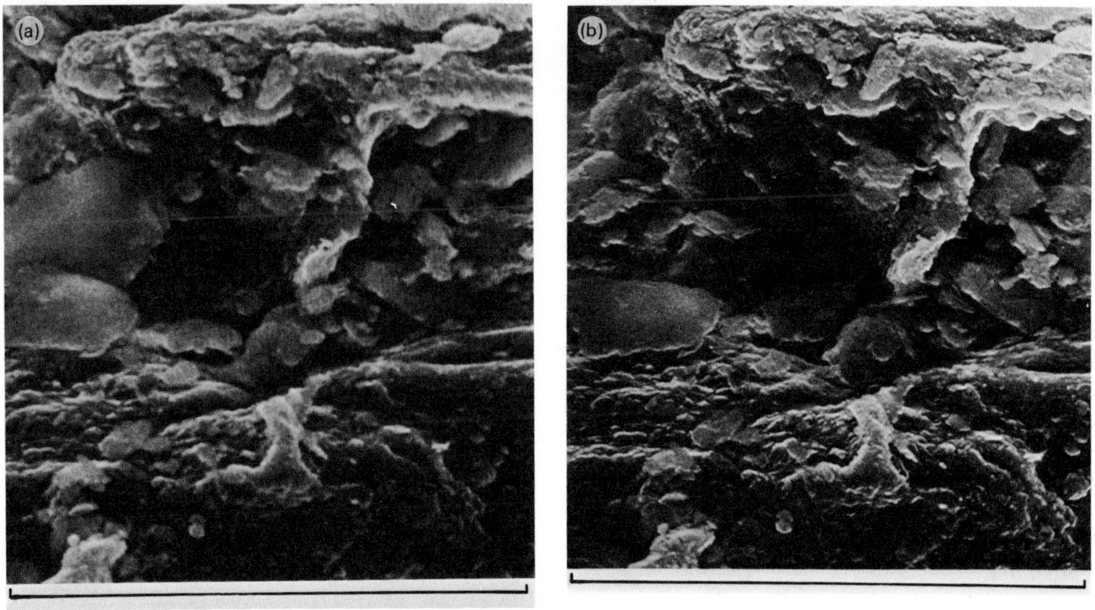

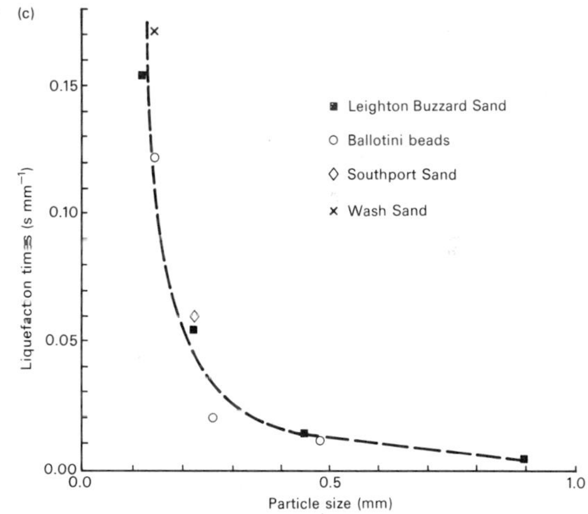

Fig. 14.12. Sand grain from Weipa Barrage, Queensland, Australia, from 8.5 m depth; 25 nm Au; SEMs, p 54 μm, θ 42°. (a) and (b) show the same area before and after 5 minutes exposure to the beam. Note the removal and reorientation of several fine particles on the surface of the sand grain; this is an example of beam damage. It was inferred that pumping or vibration might cause similar effects. Since finer particles lead to increased problems with liquefaction, see (c), use of a different borrow pit for construction was advised. Sample supplied by Binnie & Partners. 1138.40 and 42, Cambridge, bore hole J103.

15 Cathodoluminescence

In the cathodoluminescent mode, the scanning electron microscope is used to examine the light emitted from the specimen when it is irradiated by the electron beam. Cathodoluminescence can be intrinsic or extrinsic. Extrinsic cathodoluminescence arises from the composition of the sample; in quartz, the signal is controlled by minute amounts of activator or inhibitor ions included as impurities. Intrinsic cathodoluminescence arises from distortions of the crystal structure, such as by applied or residual stresses. Either the light may be analysed with a monochromator or filter to give an indication of the composition of the sample, or the intensity of light is used to form an image on the cathode-ray screen. The principal patterns seen are:

1. No cathodoluminescence, feature dark.	Presence of inhibitor ions or absence of activator ions.
2. Lack of contrast, feature uniformly bright.	Uniform distribution of impurity ions.
3. Banded structure.	Cracks healed with silica.
4. Discrete dark areas.	Crystal growth with different concentrations of impurities.
5. Diffuse dark areas.	Precipitation silica or abraded surfaces.
6. Mottled textures.	Fracture surfaces.

The examples here are of the pictorial type, taken, as is most common, in reflection. An example of transmission cathodoluminescence is given in *Techniques* Fig. 1.30; and examples of banded structures in *Techniques* Figs. 1.28 and 1.29. Throughout this Section: CL means cathodoluminescent-mode scanning electron micrograph, and EM means emissive-mode scanning electron micrograph.

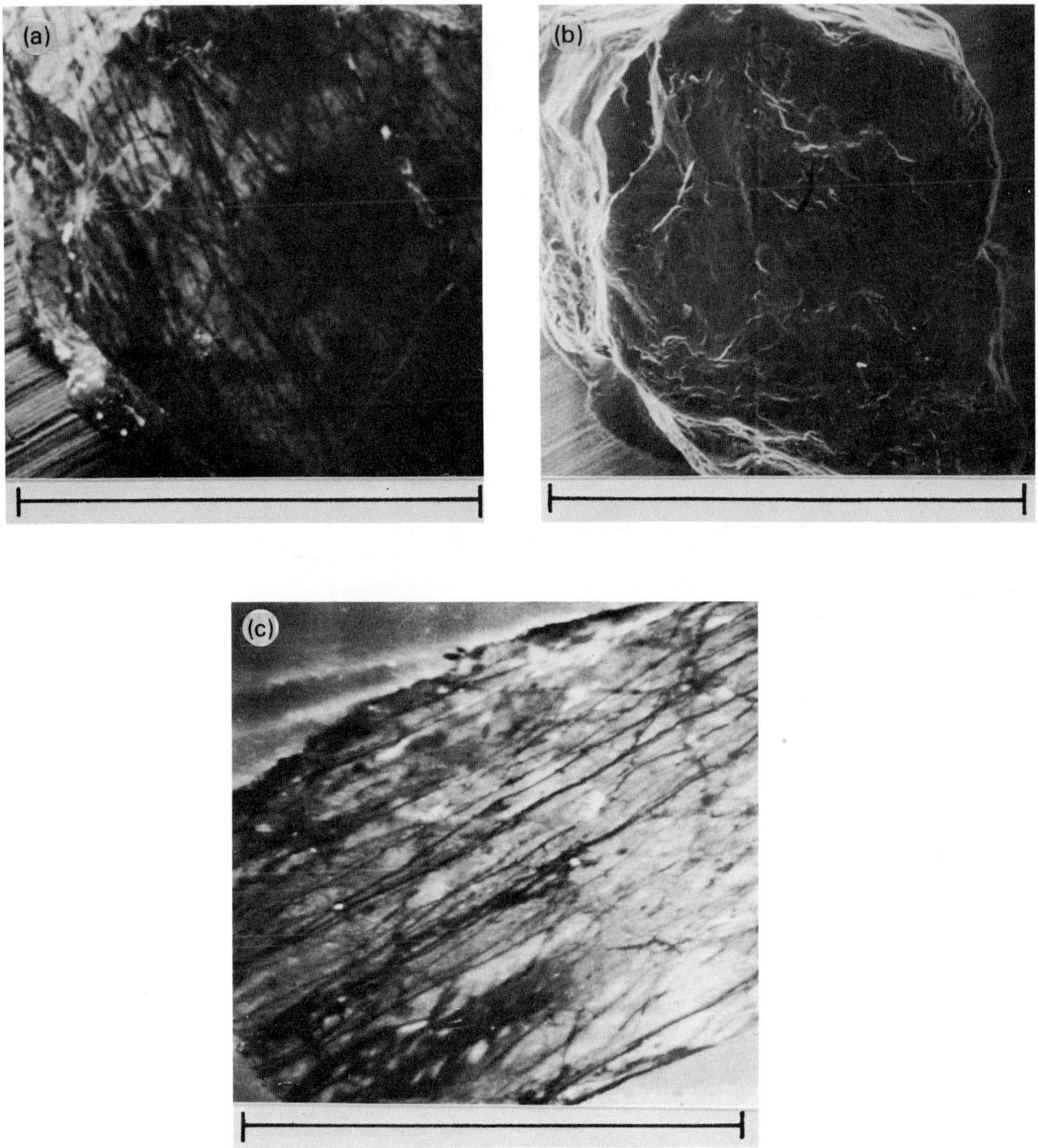

Fig. 15.1. Sand grains; (a) and (b) entire, (c) cross-section; 20 nm C; (a) and (c) CL, (b) EM, p 840, 860, 410 µm respectively, θ 42, 42, $\approx$ 50° respectively. (a) The oriented bands are evidence of metamorphic stressing; the two light bands are cracks healed in a different episode from the other bands. (b) The surficial features bear little resemblance to the bands. (c) The bands penetrate the whole grain. Details: (a) and (b) from Mimshall Wood, Hertfordshire, (c) from Libyan Desert dune. ((a) and (b) shown at a meeting of the Royal Microscopical Society, Edinburgh, 1979; (c) from Krinsley and Tovey 1978.) 618.24A, 621.12A, and 499.13A, Cambridge.

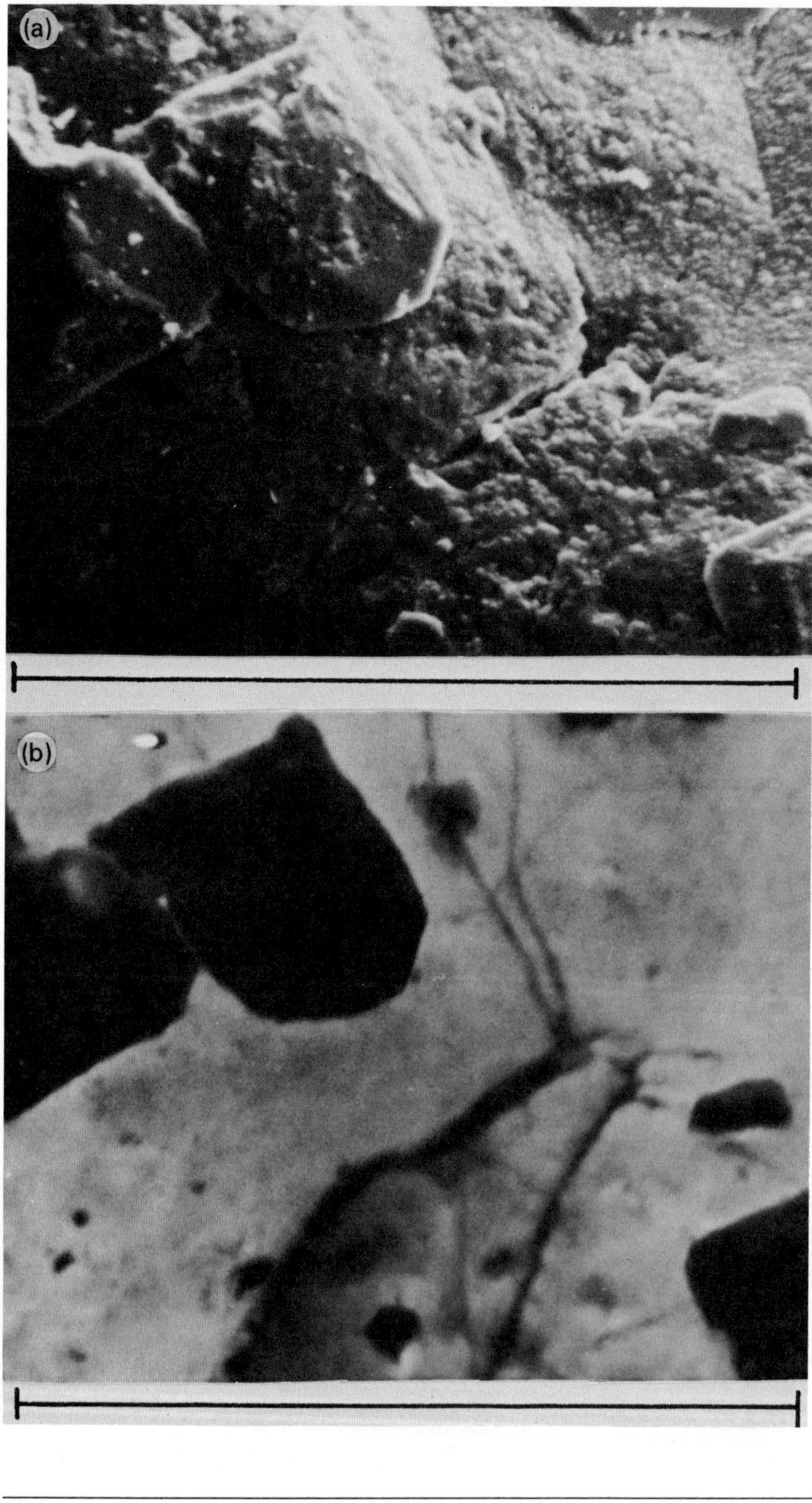
(a)
(b)

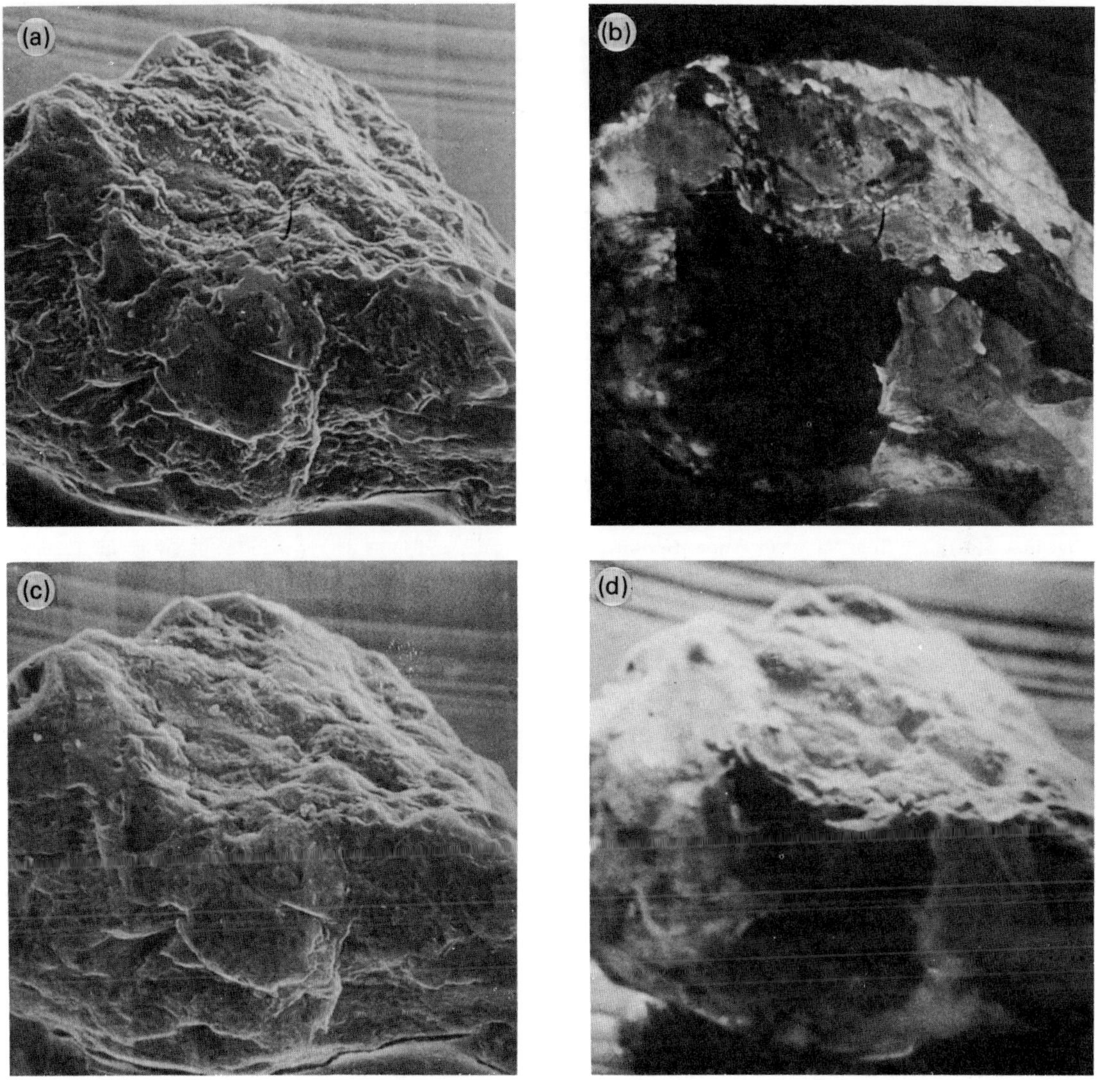

◁ **Fig. 15.2.** Sand grain from basal Permian Sandstone near Durham; 20 nm C; EM and CL respectively, p 740 μm, θ 45°. CL shows distinct boundaries, which correspond with crystal growths seen in EM; the difference in contrast indicates formation in a different environment from the parent grain. (Shown at a meeting of the Royal Microscopical Society, Edinburgh, 1979.) N. K. T., Arizona State University.

Fig. 15.3. Sand grain from Mimshall Wood, Hertfordshire; mounted individually with Araldite; (a) and (b) natural, (c) and (d) etched with 40 per cent HF for 40 minutes; 20 nm C; (a) and (c) EM-s, (c) and (d) CL-s, p 120 μm, θ 42°. (b) shows both sharply defined dark areas and 'diffuse' dark areas. The former correspond with the crystal growths in (a). After etching, most of the diffuse dark areas have been removed; however, the topographic features indicate that the thickness of the layer removed is small. These diffuse dark layers are associated with a thin amorphous deposit of silica. (Shown at a meeting of the Royal Microscopical Society, Edinburgh, 1979.) 638.38, 846.39, 852.37A, and 849.22, Cambridge.

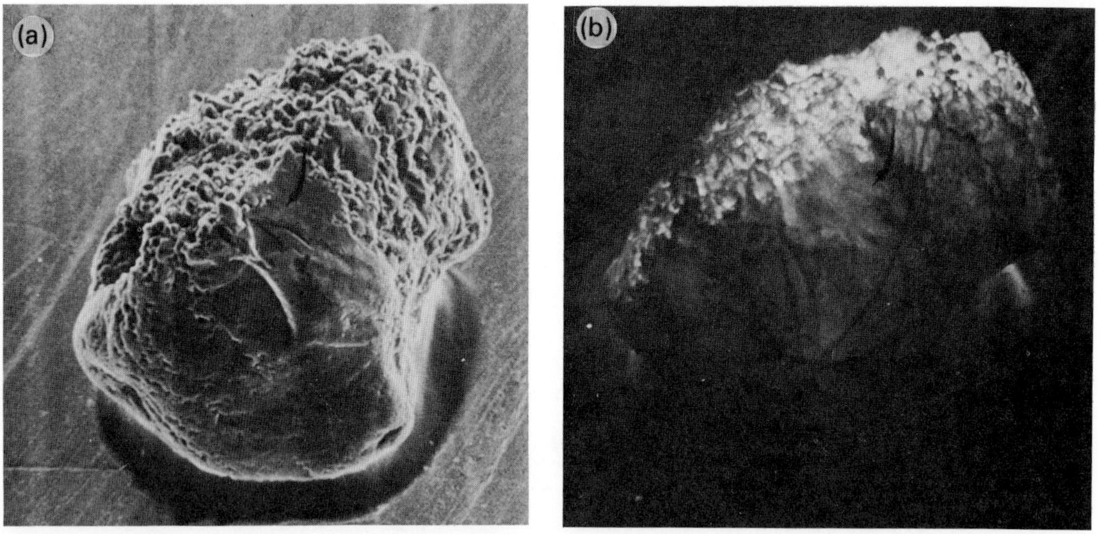

Fig. 15.4. Sand grain from Mimshall Wood, Hertfordshire; mounted individually with Araldite; fractured along upper surface; 20 nm C; EM and CL respectively, *p* 694, 670 μm respectively, θ 42°. Note the mottled texture in (b) the coincidence of many of these features with the fracture plates in (a), and that some of these features are brighter and some darker than the parent grain. This is an example of intrinsic cathodoluminescence owing to the residual stresses caused by fracturing. The radiating dark lines on the lower part of the grain in (b) suggest a point impact. (Shown at a meeting of the Royal Microscopical Society, Edinburgh, 1979.) 638.40 and 643.32, Cambridge.

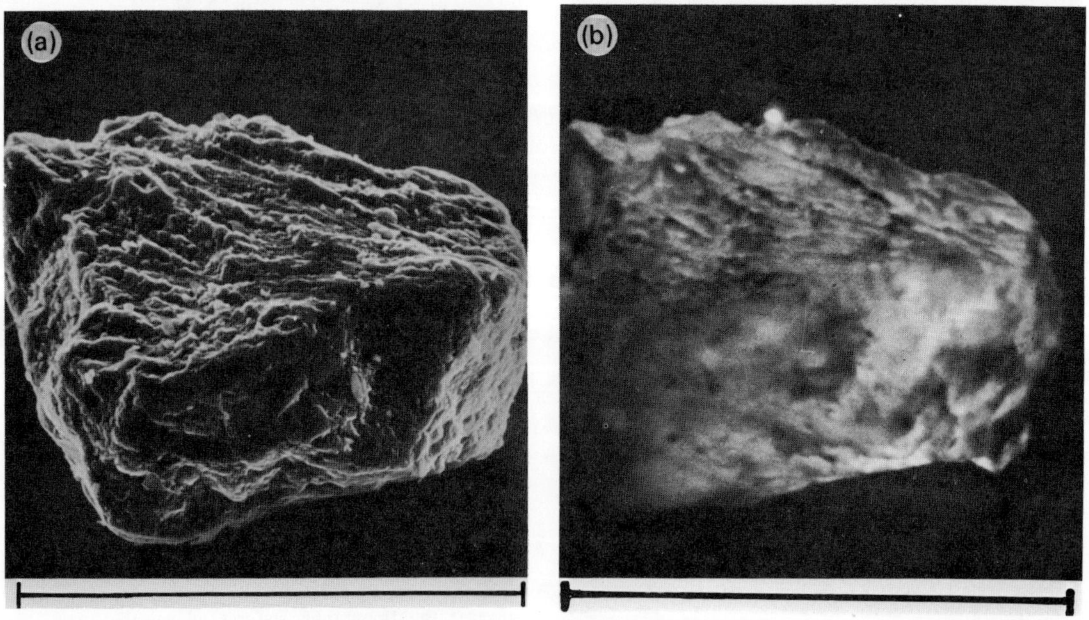

Fig. 15.5. Sand from surface near Inshore trial embankment in the Wash at TF5526; 20 nm C; EM and CL respectively, $p \approx 195\,\mu m$, θ 42°. (a) The top of the grain shows numerous plates formed by aeolian action. (b) Many CL features correspond with the general arrangement of these plates giving a naturally mottled texture. Sample supplied by P. L. Bransby. (From Krinsley and Tovey 1978.) 822.37A and 842.15, Cambridge.

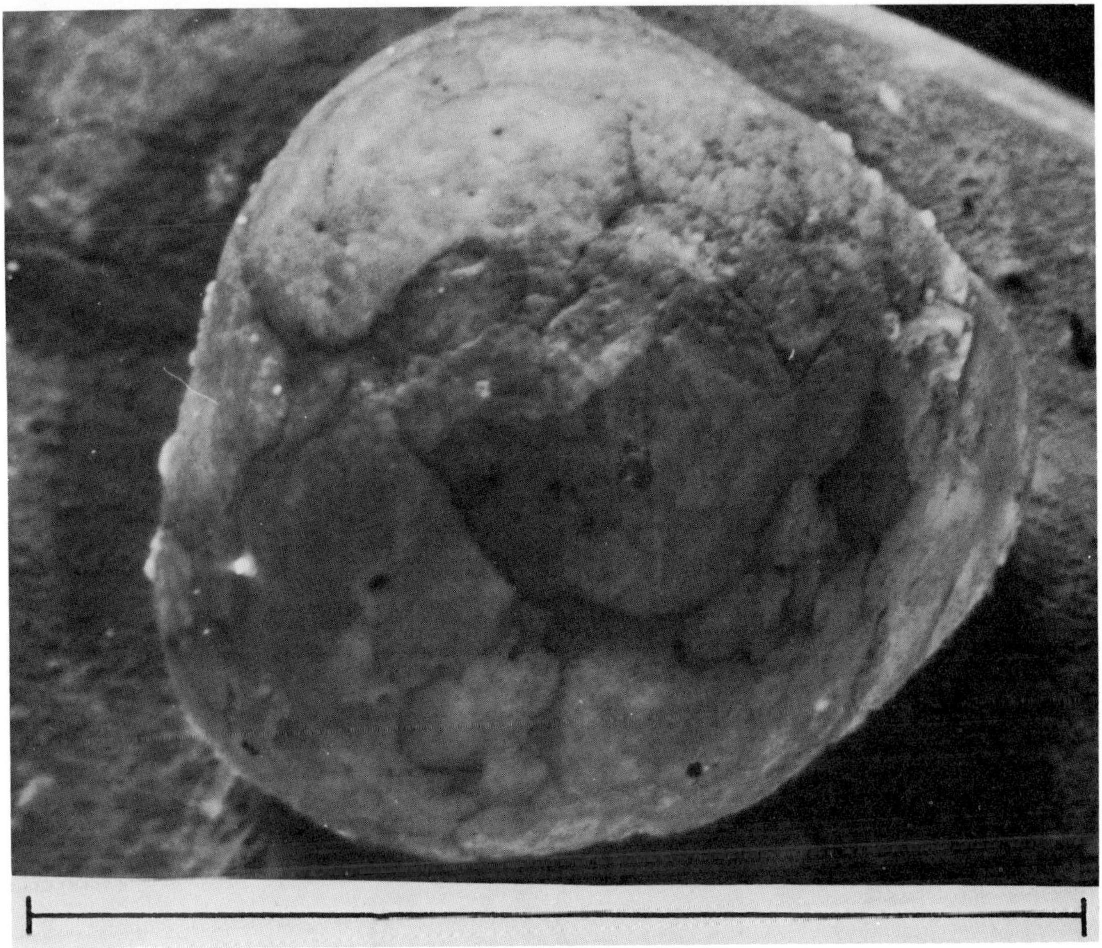

Fig. 15.6. Grain from Permian Sandstone, Durham; mounted individually with silver dag; 20 nm C; CL, *p* 1.2 mm, *θ* 45°. CL shows that the grain is an aggregate of smaller grains; evidence of this could not be seen in EM. (Shown at a meeting of the Royal Microscopical Society, Edinburgh, 1979.) 30619, N. K. T., Arizona State University.

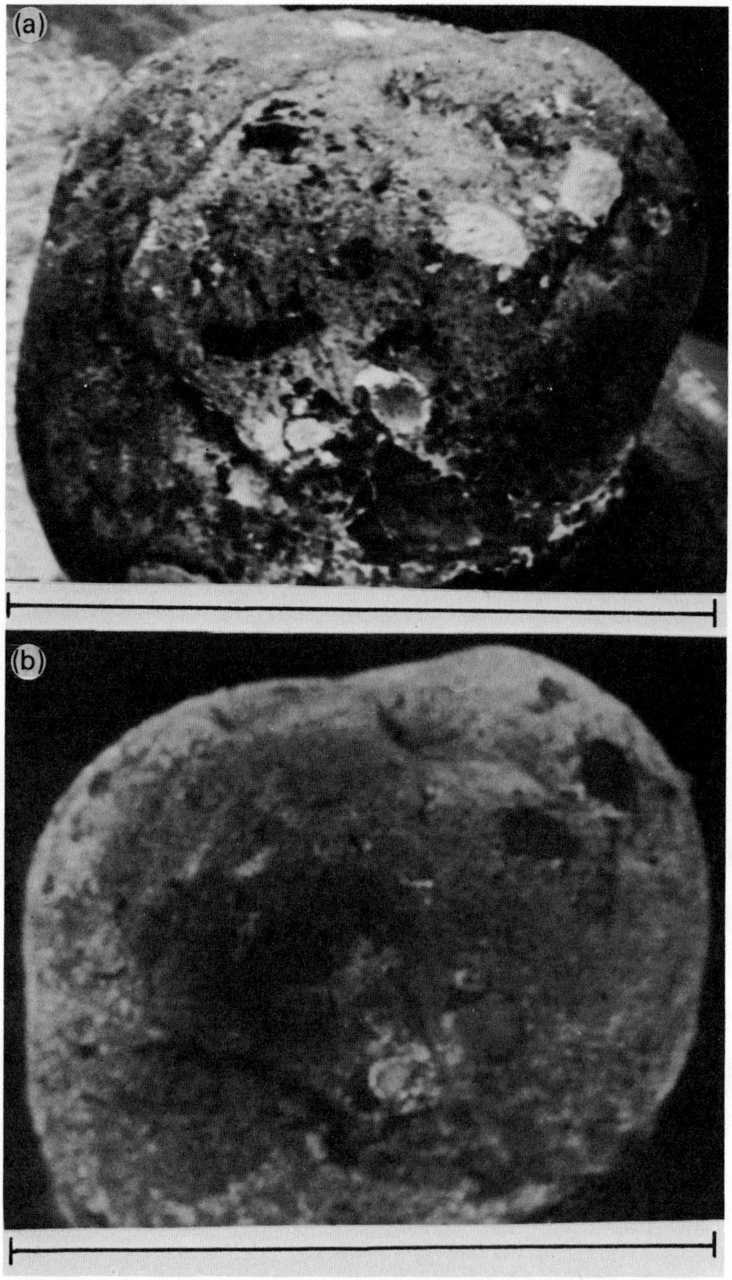

Fig. 15.7. Grain from Permian Sandstone, Durham; mounted individually with silver dag; 20 nm C; CL-s, p 170 μm, θ 45°. (a) Natural state. (b) After heating to 625 °C for 48 hours and slow cooling but not recoating. Heating has diminished the CL contrast, reversing it in some areas. A regain of contrast sometimes occurs with time, suggesfing that the features are due to trace elements, which require time to diffuse through the grain; but some of the contrast in (a) could have been due to residual stresses, which were relieved by heating. (Shown at a meeting of the Royal Microscopical Society, Edinburgh, 1979.) 30626 and 31313, N. K. T., Arizona State University.

16 Miscellaneous

A few micrographs which did not fit easily into the other Sections are included here.

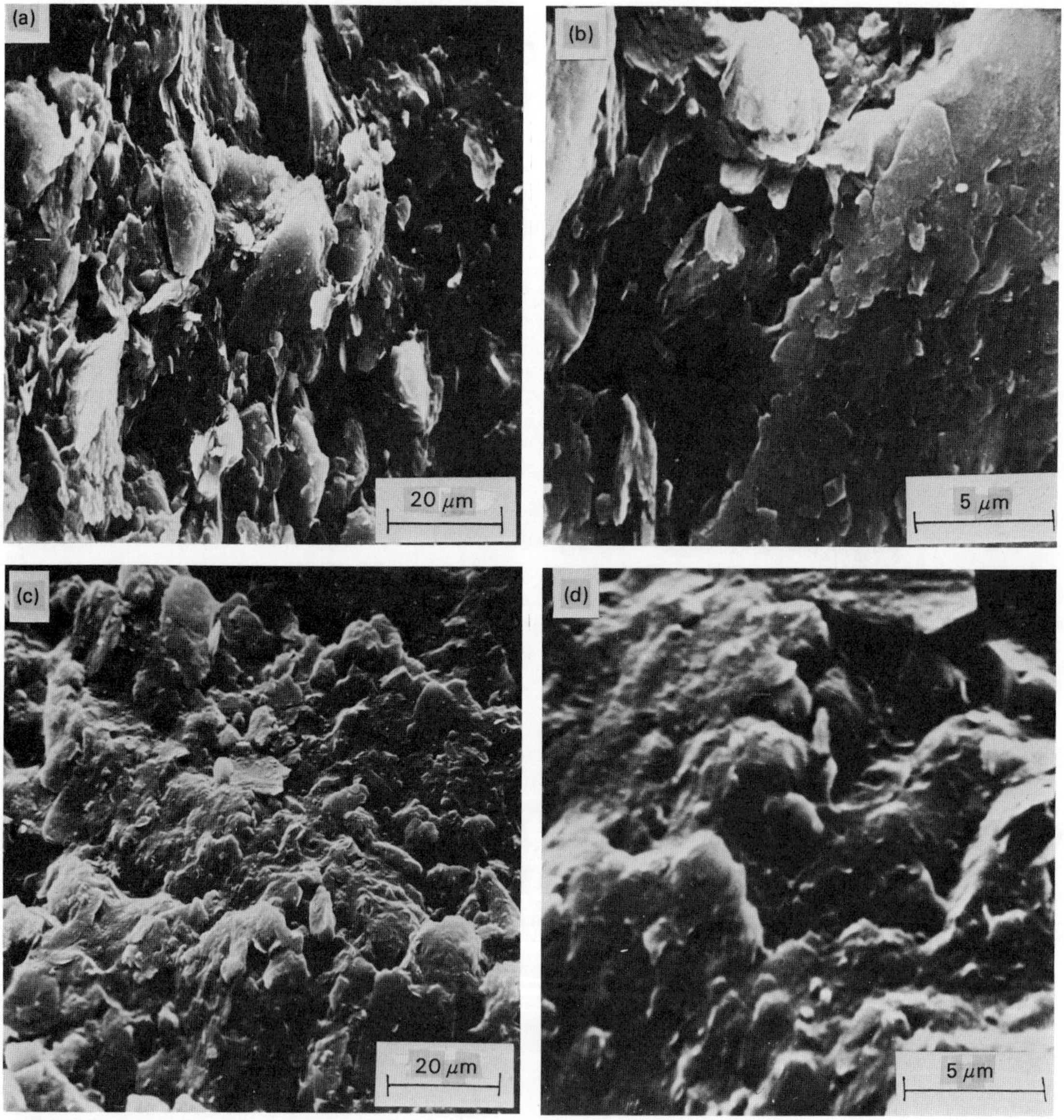

Fig. 16.1. (a) and (b). Red–brown material, (c) and (d) grey gleyed material, from the Bg or Cg horizons of Tipperty Series soil; fractured; freeze-dried; mounted on stubs; photographed in colour; coated with 10 nm C, 20 nm Au–Pd; SEMs taken using the colour photographs as guides. The red–brown material had a few pores and coatings and packets of distinct clay plates forming bridges between occasional larger particles. The gleyed material had no pores, and the clay 'plates' are indistinct. Supplied by W. J. McHardy. (From McHardy and Birnie 1975.) A. C. Birnie, Macaulay Institute.

Fig. 16.2(a). Drammen Silty Plastic Clay, 7.45 m depth; undisturbed, NGI sampler; acetone, Vestopal W; thin section ground; optical micrograph between crossed polars, which were rotated synchronously during the exposure (Leitz Dialux Pol microscope). This technique avoids a spurious impression of orientation which fixed polars may give. Optically anisotropic particles are white; voids and isotropic particles are black; clusters of small particles grey. Plates, up to 40 μm long, show a slight tendency to form a single-plate card-house framework; but the spaces within this contain smaller particles. Sample supplied by Norwegian Geotechnical Institute. 9.15, P. A. Maclean, Glasgow, 319.

Fig. 16.2(b). Drammen Clay; undisturbed; triaxial shear test; oven dried; vertical fracture; 25 nm Au; SEM, p 30 μm, θ 30°. This is from a part of the sample remote from the failure zone. Although from a different location than Fig. 16.2(a), the structure seems similar. Sample supplied by R. J. Mitchell. 18.31, Cambridge.

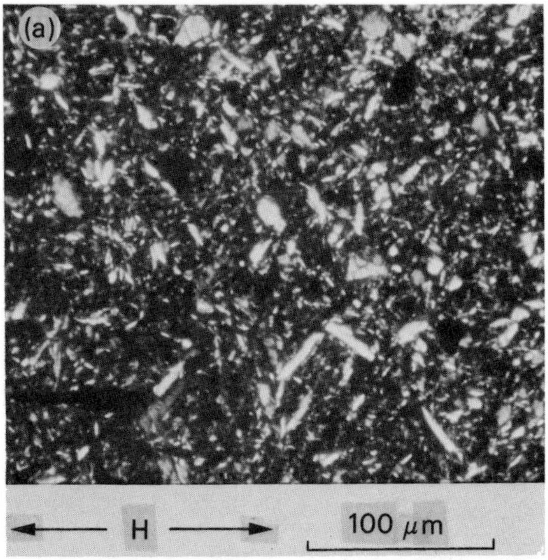

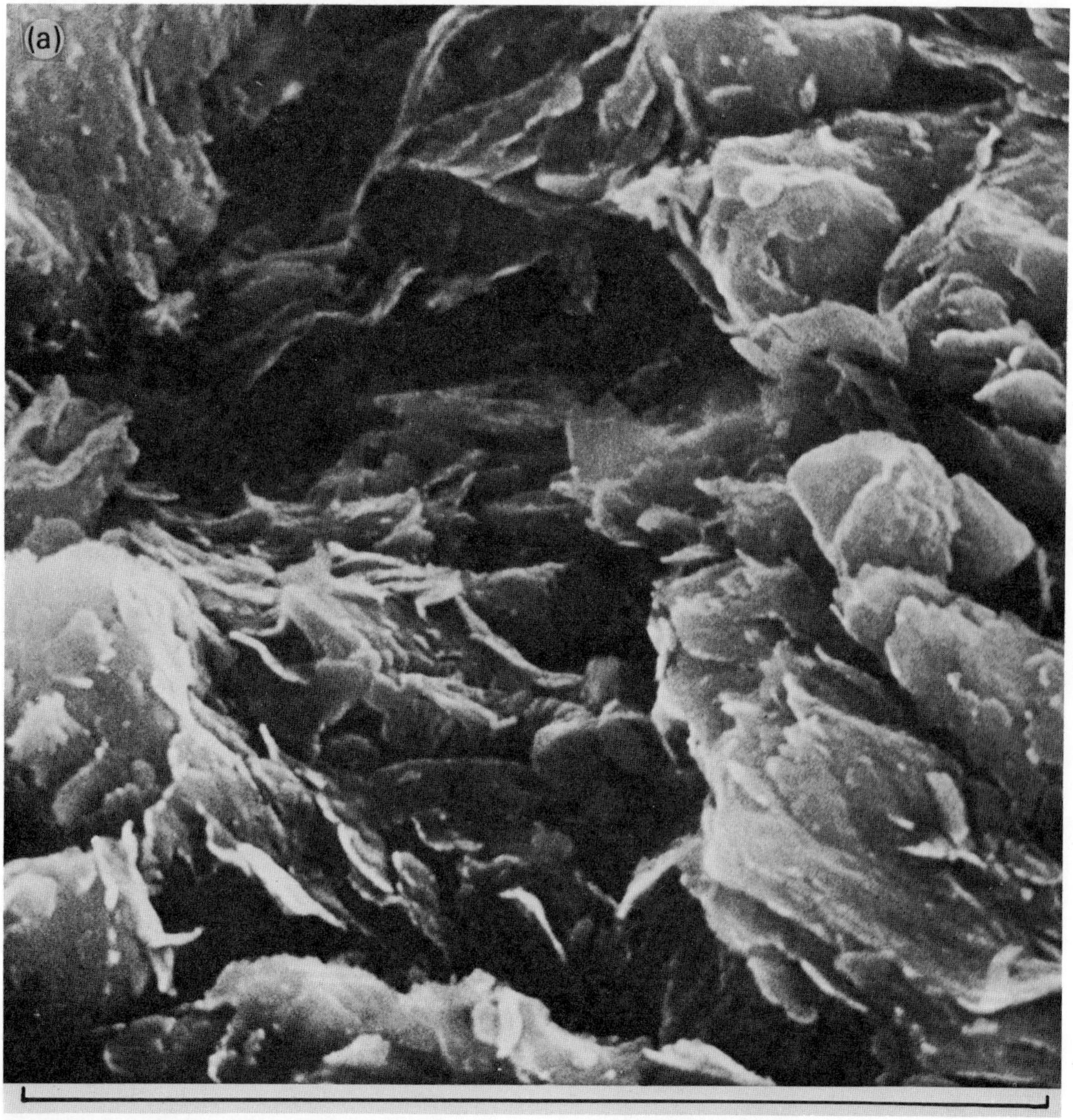

Fig. 16.3. Gault Clay from Cambridge; undisturbed; air-dried; 25 nm Au; SEMs, p 15 μm, θ 60°. Note the coccolith in (b). 126.0 and 134.31, Cambridge.

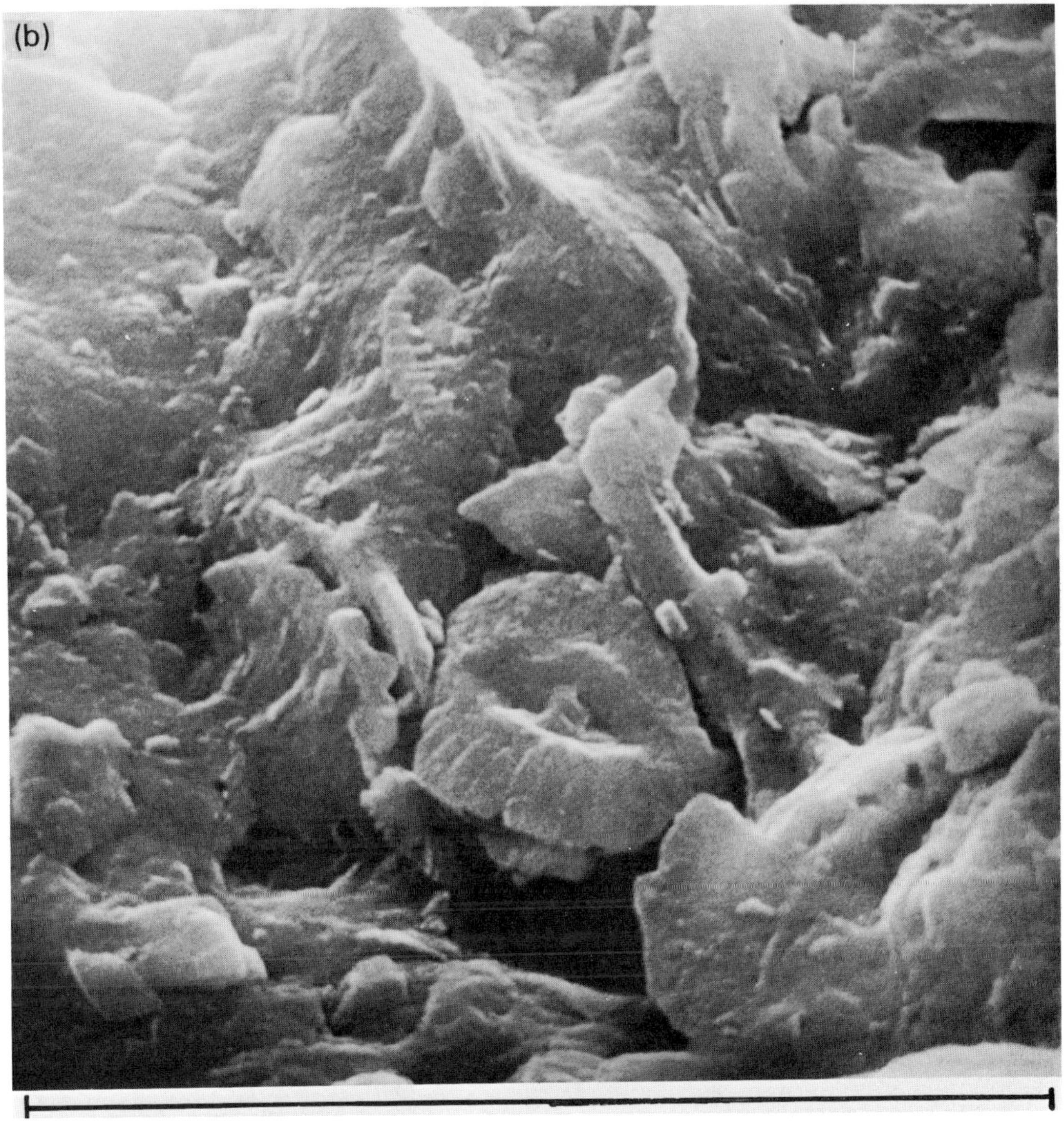
(b)

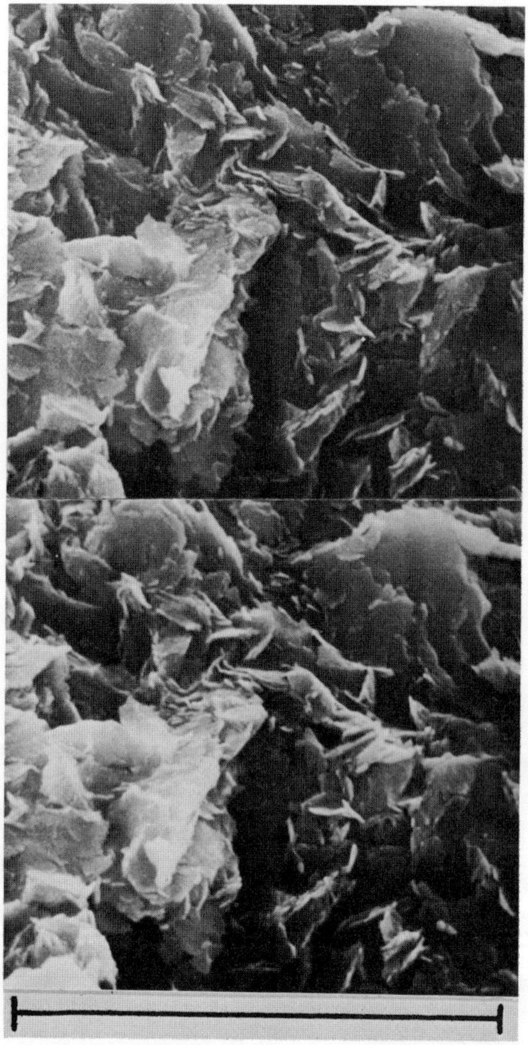

Fig. 16.4. Weald Clay; undisturbed; air-dried; vertical fracture; 25 nm Au; stereoscopic SEMs, p 15 μm, θ 40/30°. Sample supplied by E. C. Hambly. 279.18 and 279.15, Cambridge.

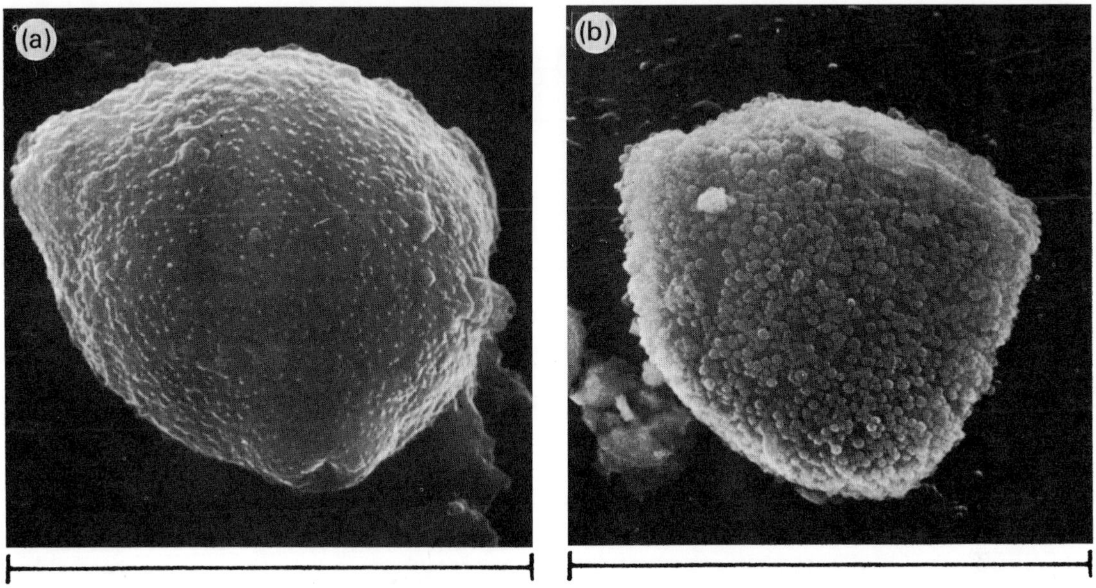

Fig. 16.5. Pollen grains from Tilbury Clay; sample air-dried; grains extracted by method of Faegri and Iversen (1966); 20 nm Au; SEMs, p 30, 23 μm respectively, θ 45°. (a) Hazel, (b) yew. Sample supplied by R. H. Parry. (From Lancaster, N. J. (1969). A study of the micro-fauna and micro-flora in clays using a scanning electron microscope. BA report, University of Cambridge.) 366.13 and 368.25, Cambridge

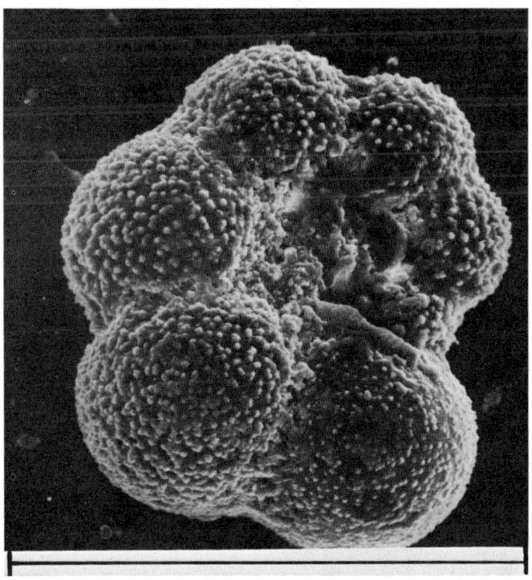

Fig. 16.6. Microfossil, globigerinelloides aspera, from Cretaceous Chalk from Caister St. Edmunds Pit, near Norwich; extracted in a solution of pyrophosphate, $NaCO_3$, and NaOH in 10 per cent H_2O_2: sample soaked overnight, boiled, sieved into three fractions using 63, 125 μm sieves; fossils selected individually; preserved in a drop of glycerine; glycerine removed with ethanol; mounted on double-sided tape using a fine brush; wetted with ethanol to orientate the fossil; 5 nm C, 20 nm Au; SEM, p 800 μm, θ 45°. Sample supplied by C. Benjamini. R. Turner, East Anglia.

17 Artefacts

Many of the standard artefacts of transmission electron microscopy are illustrated by Kay (1961, 1965); and the standard artefacts of scanning electron microscopy are described in *Techniques*, Table 6.5. This Section illustrates two artefacts which have special relevance to the electron microscopy of soils and sediments. In particular, compare the freeze-dried samples shown here with the results in Figs. 3.4–3.7.

Fig. 17.1. Anti-static spray artefact: Spestone Kaolin; remoulded at 160 per cent mc, consolidated to 200 kPa; freeze-dried; vertical fracture; sprayed with 'Durcon' anti-static spray; coated with 5 nm Au; SEM, *p* 17 μm, *θ* 42°. The sample had been observed without the gold coating, but the signal-to-noise ratio at high magnifications was poor. After applying the gold, an adequate signal-to-noise ratio was achieved. The frothy material on some of the particles was only seen on samples sprayed with the anti-static spray. No artefacts were seen at low magnification. 201.33, Cambridge.

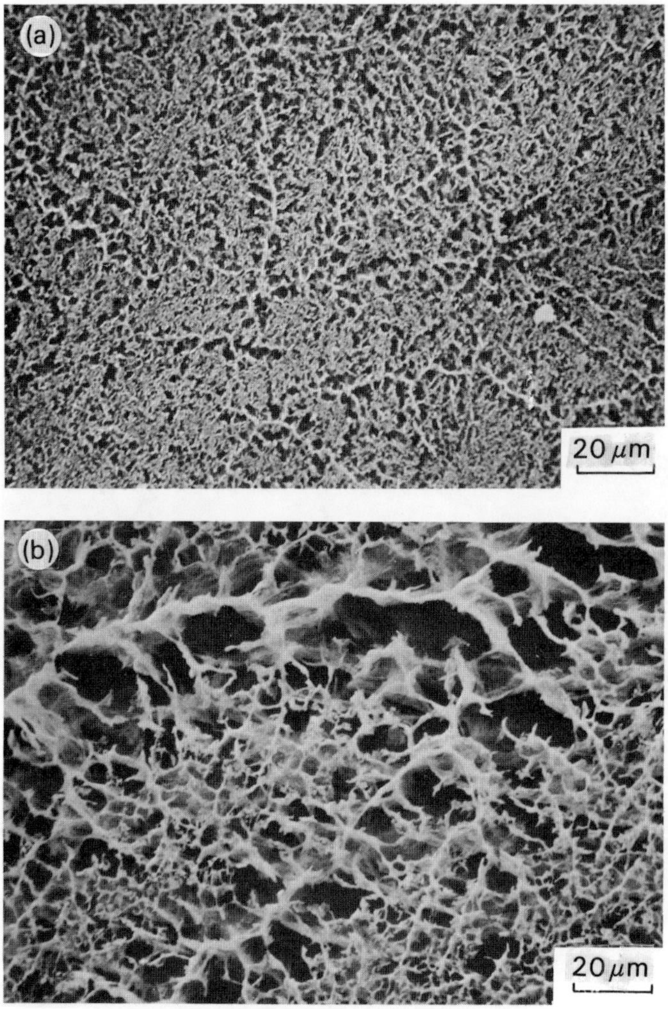

Fig. 17.2. From an evaluation of freeze-drying: Wyoming Bentonite; purified; particles larger than 30 nm nominal diameter removed; saturated with NaCl; washed free from Cl; dried at 40 °C; ground; mixed with distilled water at 76.9 g l^{-1}; (a) frozen rapidly by plunging into isopentane at 113 K; (b) frozen more slowly; sublimed at 213 K at 0.0002–0.0005 torr; 20–30 nm of C and Au; SEMs of the surfaces of the samples. Faster freezing has reduced but not eliminated structural disturbance. Supplied by R. A. Lohnes. (From Erol, Lohnes, and Demirel 1976.) Iowa State University.

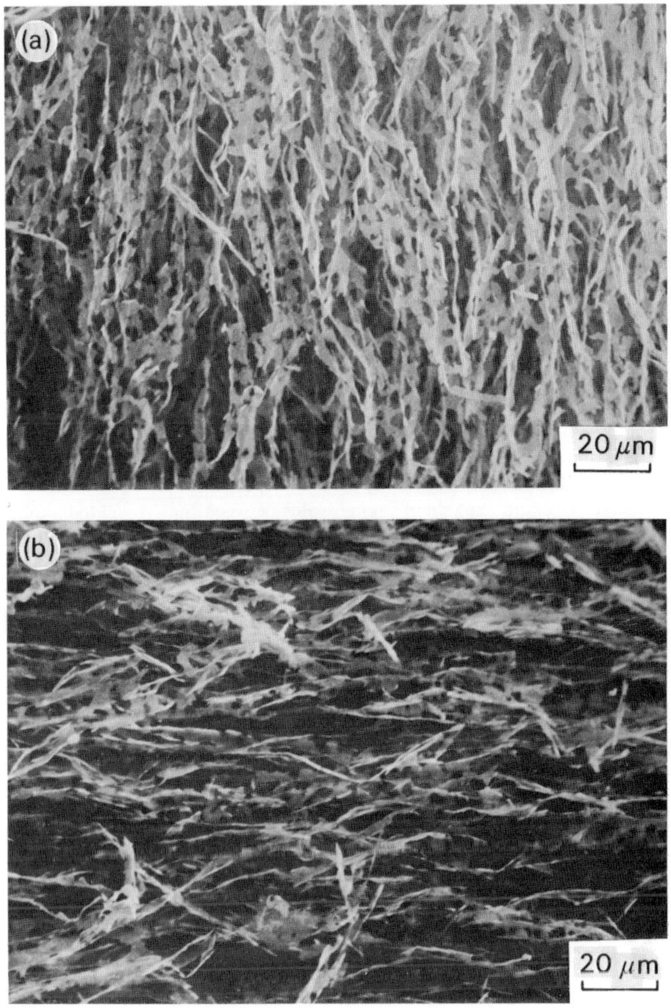

Fig. 17.3. Same as the preceding figure, except: mixed with distilled water at 7.69 g l^{-1}; fracture surfaces.

Appendix 1
Notes on terms

There is no agreed system of terminology; these notes give our current preferences. For brevity, idealized definitions are given (based on Smart 1969); these should be interpreted with latitude.

Stack: congruent plates in permanent conformal face-to-face contact, i.e. a type of particle.
Packet: a small group of plates face-to-face and capable of taking water between the plates, i.e. a very small overlap domain, see below.
Quasi-crystal: smectite flakes, say 1, 2, or 3 nm thick, face-to-face and capable of taking water between the flakes.
Cluster: a group of particles, which has a random or almost random internal structure, and possibly an anisotropic external shape; globular clusters and stepped clusters are quite different.
Domain: a group of plates parallel to a surface which may be curved; subdivided thus:
Overlap domain: the plates are separate particles;
Intergrown domain: the crystal lattices are continuous from plate to plate, perhaps imperfectly.
Bundle: a group of fibres, rods, or tubes; the principal subdivisions are into parallel, sub-parallel, and radiating bundles; radiating bundles are subdivided into sparse (i.e. sparsely filled) and close packed (i.e. more densely filled), cf. spherulite.
Spherulite: completely filled radiating 'bundle'.
Globular cluster: roughly spherical clusters of particles; subdivided principally according as they are composed of radial or circumferential plates.
Stepped cluster: a sheet of plates with face-to-face contacts with a relatively small overlap between adjacent plates; there seems to be some confusion as to whether successive plates may be joined randomly to either face of the last plate or whether they must always be on the same side like the tiles on a roof.

Cement: material which bonds particles together.
Mortar: material which fills voids without bonding to the particles.
Bridge: a long thin connection of cement or clay between (large) particles.
Buttress: a similar short stout connection.

Random structure: the particles and voids are arranged at random; often close-packed, a reasonable proportion of face-to-face plates being permitted within this class.
Single-plate card-house structure: plates arranged singly with angular or perpendicular edge-to-face contacts governed by some definite phenomena; subdivided principally into flocculated, cemented, and authigenic versions.
Book-house structure: small packets arranged with angular or perpendicular edge-to-face contacts.
Brownian structure: relatively large plates lying horizontally in a matrix of smaller particles, which themselves form any relatively unoriented and relatively open type of structure.
Honeycomb structure: particles grouped in sheets or chains, which are themselves grouped into a loose interconnecting framework with relatively large voids between the sheets.
Flocculent structure: particles grouped in loose flocs, which are themselves grouped loosely, i.e. there are small intra-floc voids and larger inter-floc voids.
Pusch's structure: clusters and loose flocs all grouped loosely with relatively large voids between these units.
Cluster structure: clusters with definite inter-cluster voids; subdivided according as anisotropy, etc., is evident.
Turbostratic structure: large domains, sometimes curved, inter-domain boundaries may be diffuse, few or no large voids.
Discrete domain structure: small domains, sometimes too small to be curved, well-defined domain boundaries, definite inter-domain voids.
cpo-structure or structure with complete preferred orientation: plates, fibres, etc., all parallel to a single direction; subdivided according as discrete domains or larger voids or neither are visible.
Tactoid structure: the original idea was of a very large quasi-crystal with most of the flakes say 1 nm apart and with a few isolated swollen voids within it; this idea may need refinement.
Dispersed structure: strictly any structure in which mutual repulsions between the particles control the structure; usually applied somewhat thoughtlessly to turbostratic or cpo-structures, either of which could be intergrown or cemented.
Herring-bone structure: plates, etc., oriented to give a zig-zag pattern.

Single-grain structure: a structure formed by clean silt or sand particles.
Granular arrangement: a relatively large occurrence of single-grain structure.

Clay arrangement: a relatively large occurrence entirely of clay.
Clay-granular arrangement: a relatively large occurrence of clay mixed with silt or sand particles.
Porphyropectic structure: a structure in which the larger particles present clean faces in fracture surfaces, and which is assumed to be uncemented.
Porphyropeptic structure: the cemented complement, the larger particles having unclean faces.

Cutan: a coating around or beside another feature.

Appendix 2
Addresses

Cambridge: Cambridge University Engineering Laboratory, Trumpington, Street, Cambridge CB2 1PZ, England.

East Anglia: School of Environmental Sciences, University of East Anglia, Norwich NR4 7TJ, England.

Glasgow: Engineering Laboratories, University of Glasgow, Glasgow G12 8QQ, Scotland.

Aylmore, Professor L. A. G., Department of Soil Science and Plant Nutrition, University of Western Australia, Nedlands, Western Australia 6009.

Bárdossy, Dr G., Research Engineering and Prime Contract Centre, Hungarian Aluminium Corporation, Budapest, Hungary.

Benjamini, Dr C., Department of Geology, Hebrew University, Jerusalem, Israel.

Bennett, Dr R. H., National Oceanic and Atmospheric Administration, Atlantic Oceanographic and Meteorological Laboratories, 15 Rickenbacker Causeway, Miami, Florida 33149, USA.

Bloomfield, Dr C., Rothamsted Experimental Station, Harpenden, Hertfordshire AL5 2JQ, England.

Climpson, Mr N. A., English Clays Lovering Pochin & Co. Ltd., John Keay House, St. Austell, Cornwall PL25 4DJ, England.

Crawford, Dr C. B., NRC Division of Building Research, Ottawa K1A OR6 Canada.

Darbyshire, Dr J. F., Macaulay Institute for Soil Research, Craigiebuckler, Aberdeen AB9 2QJ, Scotland.

Defence Standards Laboratory, PO Box 50, Ascot Vale W.2, Victoria, Australia.

Diamond, Professor S., School of Civil Engineering, Purdue University, West Lafayette, Indiana 47907, USA.

FitzPatrick, Dr E. A., Department of Soil Science, University of Aberdeen, Meston Walk, Aberdeen AB9 2UE, Scotland.

Foster, Dr R. C., CSIRO Division of Soils, Private Bag No 2, Glen Osmond, S.A. 5064, Australia.

Gillott, Professor J. E., Department of Civil Engineering, University of Calgary, 2920 24 Ave NW, Calgary T2N 1N4, Canada.

Keller, Professor W.D., Department of Geology, University of Missouri, Columbia, Missouri 65201, USA.

Kirkman, Professor J. H., Department of Soil Science, Massey University, Palmerston North, New Zealand.

Kodama, Dr H., Chemistry and Biology Research Institute, Agriculture Canada, Ottawa K1A OC6, Canada.

Krinsley, Professor D. H., Department of Geology, Arizona State University, Tempe Arizona 85281, USA.

Lohnes, Professor R. A., Department of Civil Engineering, Iowa State University, Ames, Iowa 50011, USA.

Low, Mrs A. J., Department of Soil Science, University of Aberdeen, Meston Walk, Aberdeen AB9 2UE, Scotland.

McConnochie, Dr I., Department of Civil Engineering, University of Glasgow, Glasgow G12 8LT, Scotland.

McHardy, Dr W. J., Macaulay Institute for Soil Research, Craigiebuckler, Aberdeen AB9 2QJ, Scotland.

Marshall, Professor K. C., School of Microbiology, University of New South Wales, PO Box 1, Kensington, New South Wales 2033, Australia.

Mattiat, Dr B., Niedersächsisches Landesamt für Bodenforschung, 3 Hannover 23, Postfach 23 01 53, West Germany.

Mitchell, Professor J. K., Department of Civil Engineering, University of California, Berkeley, California 94720, USA.

Noble, Mr F.R., English Clays Lovering Pochin & Co. Ltd., John Keay House, St. Austell, Cornwall PL25 4DJ, England.

O'Brien, Professor N. R., Department of Geological Sciences, State University College, Potsdam, New York 13676, USA.

Oertel, Professor G., Department of Earth and Space Sciences, University of California, Los Angeles, California 90024, USA.

Olphen, Dr H van, Dalweg 114, 6865 CW Doorwerth, The Netherlands.

Paulus, Professor M., Laboratoire d'Étude et de Synthèse des Microstructures, CNRS-ESPCI, 10 Rue Vauquelin, 75231 Paris Cédex 05, France.

Penner, Dr E., NRC Division of Building Research, Ottawa K1A OR6, Canada.

Prichard, Miss H. M., Department of Earth Sciences, Open University, Milton Keynes, Buckinghamshire MK7 6AA, England.

Pusch, Professor R., Division of Soil Mechanics, Högskolan i Luleå, 951 87 Luleå, Sweden.

Quigley, Professor R. M., Civil Engineering Group, University of Western Ontario, London, Ontario, Canada.

Rautureau, Dr M., Laboratoire de Cristallographie, Université d'Orléans, Rue de Chartres, 45045 Orléans Cédex, France.

Reichenbach, Professor Dr H. Graf von, Institut für Bodenkunde, Technischen Universität, Herrenhäuser Str. 2, 3 Hannover 21, West Germany.

Silva, Mr S. R., Department of Defence Materials Research Laboratories, PO Box 50, Ascot Vale, Vic. 3032, Australia.

Tazaki, Dr Kazue, Institute for Thermal Spring Research, Okayama University, Misasa, Tottori-Ken 682–02, Japan.

Tomlinson, Mr, M. J., Wimpey Ltd, Hayes, Middlesex.

Weir, Dr A. H., Rothamsted Experimental Station, Harpenden Hertfordshire AL5 2JQ, England.

West, Professor R., State College of Ceramics, Alfred University, Alfred, New York 14802, USA.

Whalley, Dr W. B., Department of Geography, The Queen's University, Belfast BT7 1NN, Northern Ireland.

Yaalon, Professor D. H., Department of Geology, Hebrew University, Jerusalem, Israel.

Yada, Professor K., Research Institute for Scientific Measurements, Tohoku University, Sanjyo-Machi, Sendai, Japan.

Yong, Professor R. N., Geotechnical Research Centre, McGill University, 817 Sherbrooke St. W., Montreal, PQ, H3A 2K6, Canada.

References

Allpress, J. G. and Sanders, J. V. (1973). The direct observation of the structure of real crystals by lattice imaging. *J. appl. Crystallogr.* **6,** 165–90.

Aylmore, L. A. G. and Quirk, J. P. (1959). Swelling of clay-water systems. *Nature, Lond.* **183,** 1752–3.

—— —— —— —— (1960*a*). Domain or turbostratic structure of clays. *Nature, Lond.* **187,** 1046–8.

—— —— —— —— (1960*b*). The structural status of clay systems. *Proc. 9th Nat. Conf. Clays Clay Miner.*, pp. 104–30.

Bárdossy, G., Csanády, A., and Csordás, A. (1978). Scanning electron microscope study of bauxites of different ages and origins. *Clays Clay Miner.* **26,** 245–62.

Bennett, R. H., Bryant, W. R., and Keller, G. H. (1977). Clay fabric and geotechnical properties of selected submarine sediment cores from the Mississippi Delta. Professional Paper 9, NOAA Atlantic Oceanographic and Meteorological Laboratories, Miami, Florida.

Beutelspacher, H. and van der Marel, H. W. (1968). *Atlas of electron microscopy of clay minerals and their admixtures.* Elsevier, Amsterdam.

Brown, G. and Weir, A. H. (1963). The identity of rectorite and allevardite. *Proc. 1st Int. Clay Conf.*, pp. 27–35.

Darbyshire, J. F. and Greaves, M. P. (1973). Bacteria and protozoa in the rhizosphere. *Pestic. Sci.* **4,** 349–60.

Diamond, S. and Bloor, J. W. (1970). Globular cluster microstructure of endellite (hydrated halloysite) from Bedford, Indiana. *Clays Clay Miner.* **18,** 309–12.

Erol, O., Lohnes, R. A., and Demirel, T. (1976). Preparation of clay-type moisture-containing samples for scanning electron microscopy. In *Scanning electron microscopy 1976* (ed. O. Johari) pp. 769–76. SEM Inc., O'Hare, Illinois.

Faegri, R., and Iversen, J. (1966). *Text book of pollen analysis.* Blackwell, Oxford.

Farmer, V. C., Russell, J. D., McHardy, W. J., Newman, A. C. D., Ahlrichs, J. L., and Rimsaite, J. Y. H. (1971). Evidence for loss of protons and octahedral iron from oxidized biotites and vermiculites. *Mineralog. Mag.* **38,** 121–37.

Foster, R. H. and De, P. K. (1971). Optical and electron microscopic investigation of shear induced structures in lightly consolidated (soft) and heavily consolidated (hard) kaolinite. *Clays Clay Miner.* **19,** 31–47.

Gard, J. A. (ed.) (1971). *The electron-optical investigation of clays.* Mineralogical Society, London.

Greaves, M. P. and Darbyshire, J. F. (1972). The ultrastructure of the mucilaginous layer on plant roots. *Soil Biol. Biochem.* **4,** 443–9.

Kay, D. (1961). *Techniques for electron microscopy* (2nd edn 1965). Blackwell, Oxford.

Keller, W. D. (1968). Flint clay and a flint-clay facies. *Clays Clay Miner.* **16,** 113–28.

—— (1978). Classification of kaolins exemplified by their textures in scan electron micrographs. *Clays Clay Miner.* **26,** 1–20.

King, A. G. (1960). The influence of microstructure on the mechanical properties of dense polycrystalline alumina. In *Mechanical properties of engineering ceramics* (ed. W. W. Kriegel and H. Palmour), pp. 333–47. Interscience, New York.

Kirkman, J. H. (1977). Possible structure of halloysite disks and cylinders observed in some New Zealand rhyolitic tephras. *Clay Miner.* **12,** 199–216.

Kodama, H. (1977). An electron-diffraction study of a microcrystalline muscovite and its vermiculitized products. *Mineralog. Mag.* **41,** 461–8.

Krinsley, D. H. and Doornkamp, J. C. (1973). *Atlas of quartz sand surface textures.* Cambridge University Press.

—— and Tovey, N. K. (1978). Cathodoluminescence in quartz sand grains. In *Scanning electron microscopy 1978* (ed. O. Johari), Vol 1, pp. 887–94. SEM Inc., O'Hare, Illinois.

Kubiëna, W. L. (1938). *Micropedology.* Collegiate Press, Ames, Iowa.

McConnachie, I. (1974). Fabric changes in consolidated kaolin. *Geotechnique* **24,** 207–22.

McHardy, W. J. (1977). Application of scanning electron microscopy and microanalysis to the study of soil mineralogy. *Geol. Appl. Idrogeol.* **12,** 213–33.

—— and Birnie, A. C. (1975). Scanning electron microscope studies of a surface water gley. *J. Soil Sci.* **26,** 426–31.

McKyes, E., Sethi, A., and Yong, R. N. (1974). Amorphous coatings on particles of sensitive clay soils. *Clays Clay Miner.* **22,** 427–33.

Marshall, K. C. (1975). Clay mineralogy in relation to survival of soil bacteria. *A. Rev. Phytopath.* **13,** 357–73.

Mattiat, B. (1969). Eine Methode zur elektronenmikroskopischen Untersuchung des Mikrogefüges in tonigen Sedimenten. *Geol. Jber.* **88,** 87–111.

Mitchell, J. K. and El Jack, S. A. (1965). The fabric of soil cement and its formation. *Proc. 14th Nat. Conf. Clays Clay Miner.* pp. 297–305.

Noishiki, Y. and Tazaki, K. (1975). [Applications of critical point drying for scanning electron microscopic study of clay minerals.] *Pap. Instn. therm. Spring Res. Okayama Univ.* **44,** 1–6. [Japanese with English abstract.]

O'Brien, N. R. (1970). The fabric of shale—an electron-microscope study. *Sedimentology* **15,** 229–46.
—— (1971). Fabric of kaolinite and illite floccules. *Clays Clay Miner.* **19,** 353–9.
Oertel, G. and Phakey, P. P. (1972). The texture of a slate from Nantlle, Caernarvon, North Wales. *Texture* **1,** 1–8.
van Olphen, H. (1977). *An introduction to clay colloid chemistry.* Wiley, New York.
Osipov, V. I. and Sokolov, V. N. (1978). Relation between the microfabric of clay soils and their origin and degree of compaction. *Bull. int. Ass. Engng. Geol.* **18,** 73–81.
Paulus, M., Dubon, A., and Etienne, J. (1975). Application of ion-thinning to the study of the structure of argillaceous rocks by transmission electron microscopy. *Clay Miner.* **10,** 417–26.
Penner, E., Gillott, J. E., and Eden, W. J. (1970). Investigation of heave in Billings shale by mineralogical and biogeochemical methods. *Can. Geotech. J.* **7,** 333–8.
Prichard, H. M. (1979). A petrographic study of the process of serpentinisation in ophiolites and ocean crust. *Contr. Miner. Petrol.* **68,** 231–41.
Quigley, R. M., Zajic, J. E., McKyes, E., and Yong, R. N. (1973). Biochemical alteration and heave of black shale; detailed observations and interpretations. *Can. J. earth Sci.* **10,** 1005–15.
Rautureau, M. S. and Tchoubar, C. (1976). Structural analysis of sepiolite by selected area electron diffraction—relations with physico-chemical properties. *Clays Clay Miner.* **24,** 43–9.
Reichenbach, H. Graf von (1972). Exchange equilibria of interlayer cations in different particle size fractions of biotite and phlogopite. *Proc. 4th Int. Clay Conf.*, pp. 457–66.
Roper, M. R. and Marshall, K. C. (1974). Modification of the interaction between *Escherichia coli* and bacteriophage in saline sediment. *Microb. Ecol.* **1,** 1–13.
Sergeyev, Y. M., Grabowska-Olszewska, B., Osipov, V. I., and Sokolov, V. N. (1978). Types of the microstructures of clayey soils. In *Proc. 3rd Int. Cong. IAEG*, Section 2, Vol. 1, pp. 319–27.
Smart, P. (1966). Particle arrangements in kaolin. In *Proc. 15th Nat. Conf. Clays Clay Miner.*, pp. 241–54.
—— (1969). Soil structure in the electron microscope. In *Structure, solid mechanics and engineering design—Proc. Southampton Civil Engng. Materials Conf.* (ed. M. Te'eni), Vol. 1, pp. 249–55. Wiley, London.
—— (1975). Soil microstructure. *Soil Sci.* **119,** 385–93.
—— and Dickson, J. W. (1978). Deformation and shear of normally consolidated flocculated kaolin. In *Proc. Interdiscip. Conf. Mech. Deformation Fracture*, Luleå, Vol. 1, pp. 121–8. University of Luleå, Sweden.
Tazaki, K. (1976). Scanning electron microscopic study of formation of gibbsite from plagioclase. *Pap. Instn. therm. Spring Res. Okayama Univ.* **45,** 11–24.
—— (1978*a*). [Micromorphology of plagioclase surface at incipient stage of weathering.] *Earth Sci., Toyko* **32,** 58–62. [Japanese with English abstract.]
—— (1978*b*). Micromorphology of halloysite produced by weathering of plagioclase in volcanic ash. *Proc. 6th Int. Clay Conf.*, pp. 415–22.
—— (1979). Scanning electron microscopic study of imogolite formation from plagioclase. *Clays Clay Miner.* **27,** 209–12.
——Tazaki, K., and Komatsu, M. (1978). [Clay mineralogy of volcanoclastic and sedimentary rocks from the Marinana Trench.] *J. Clay Sci. Soc. Japan* **18,** 51–62. [Japanese with English abstract.]
Thiessen, P. A. (1942). Wechselseitige Adsorption von Kolloiden. *Z. Elektrochem.* **48,** 675–81.
Tovey, N. K. (1973). Some applications of electron microscopy to soil engineering. In *Soil microscopy. Proc. 4th Int. Working Meeting Soil Micromorphology*, Kingston (ed. G. K. Rutherford), pp. 119–42. Limestone Press, Kingston, Ontario.
—— and Wong, N. K. (1977). Preparation, selection and interpretation problems in scanning electron microscope studies of sediments. In *Scanning electron microscopy in the study of sediments—a symposium* (ed. W. B. Whalley), pp. 181–99. Geo Abstracts, Norwich.
—— Frydman, S., and Wong, K. Y. (1973). A study of swelling clays in the SEM. In *Proc. 3rd Int. Conf. Expansive Soils*, Haifa, Vol. 2, pp. 89–98.
Weir, A. H. (1965). Potassium retention in montmorillonites. *Clay Miner.* **6,** 17–22.
—— Nixon, H. L., and Woods, R. D. (1960). Measurement of thickness of dispersed clay flakes with the electron microscope. *Clays Clay Miner.* **9,** 419–23.
West, R. (1963). The characteristics of filter pressed kaolinite–water pastes. *Proc. 12th Nat. Conf. Clays Clay Miner.*, pp. 209–21.
Whalley, W. B. (1977). Scanning electron microscopic examination of a laboratory-simulated silcrete. In *Scanning electron microscopy in the study of sediments—a symposium* (ed. W. B. Whalley), pp. 399–405. Geo Abstracts, Norwich.
Wieder, M. and Yaalon, D. H. (1974). Effect of matrix composition on carbonate nodule crystallization. *Geoderma* **11,** 95–121.
Wilson, M. J., Bain, D. C., and McHardy, W. J. (1971). Clay mineral formation in a deeply weathered boulder conglomerate in North-East Scotland. *Clays Clay Miner.* **19,** 345–52.
Wong, K. Y. and Tovey, N. K. (1975). A new specimen preparation technique for the scanning electron microscope. *Geotechnique* **25,** 142–5.
Yada, K. (1967). Study of chrysotile asbestos by a high resolution electron microscope. *Acta crystallogr.* **23,** 704–7.

Subject Index

T denotes a Figure number in *Techniques*. ( ) indicates implied or presumed. There is a fuller index of techniques in *Techniques*.

Name Index